MOLECULAR
BIOLOGY
INTELLIGENCE
UNIT

INTRACELLULAR SIGNAL TRANSDUCTION: THE JAK-STAT PATHWAY

Andrew F. Wilks
Ailsa G. Harpur
Ludwig Institute for Cancer Research
Melbourne, Victoria, Australia

Springer-Verlag
Berlin Heidelberg GmbH

R.G. LANDES COMPANY
AUSTIN

Molecular Biology Intelligence Unit
INTRACELLULAR SIGNAL TRANSDUCTION: THE JAK-STAT PATHWAY

R.G. LANDES COMPANY
Austin, Texas, U.S.A.

Springer

International ISBN 978-3-662-22052-8

While the authors, editors and publisher believe that drug selection and dosage and the specifications and usage of equipment and devices, as set forth in this book, are in accord with current recommendations and practice at the time of publication, they make no warranty, expressed or implied, with respect to material described in this book. In view of the ongoing research, equipment development, changes in governmental regulations and the rapid accumulation of information relating to the biomedical sciences, the reader is urged to carefully review and evaluate the information provided herein.

Library of Congress Cataloging-in-Publication Data

Wilks, Andrew F.
Intracellular Signal Transduction: The JAK-STAT Pathway / Andrew F. Wilks, Ailsa G. Harpur
p. cm. — (Molecular biology intelligence unit)
Includes bibliographical references and index.
ISBN 978-3-662-22052-8 ISBN 978-3-662-22050-4 (eBook)
DOI 10.1007/978-3-662-22050-4
1. Cellular signal transduction. 2. Protein kinases. 3. Cytokines. 4. Cytokines--Receptors. 5. Second messengers (Biochemistry) I. Harpur, Ailsa G. II. Title. III. Series.
QP517.C45W55 1996
574.87--dc20 94-9438
CIP

Publisher's Note

R.G. Landes Company publishes six book series: *Medical Intelligence Unit, Molecular Biology Intelligence Unit, Neuroscience Intelligence Unit, Tissue Engineering Intelligence Unit, Biotechnology Intelligence Unit* and *Environmental Intelligence Unit.* The authors of our books are acknowledged leaders in their fields and the topics are unique. Almost without exception, no other similar books exist on these topics.

Our goal is to publish books in important and rapidly changing areas of bioscience and environment for sophisticated researchers and clinicians. To achieve this goal, we have accelerated our publishing program to conform to the fast pace in which information grows in bioscience. Most of our books are published within 90 to 120 days of receipt of the manuscript. We would like to thank our readers for their continuing interest and welcome any comments or suggestions they may have for future books.

Shyamali Ghosh
Publications Director
R.G. Landes Company

CONTENTS

CHAPTER 1

INTRACELLULAR SIGNAL TRANSDUCTION

Intracellular signal transduction is the mechanism by which cells acquire information from their extracellular milieu and convert that information into chemical stimuli upon which they can make relevant developmental and functional decisions. For the purposes of this book, we will only address those issues which are pertinent to the intracellular signal transduction processes triggered by growth factors and cytokines. Our understanding of the molecular mechanisms that are the basis of intracellular signal transduction has burgeoned over the last twelve to fifteen years, initially due to our increased understanding of the molecular biology of viral oncogenesis, and more recently through a concerted effort to understand the structure/function relationships of those molecules that are involved in signal transduction. Intracellular signal transduction mechanisms underpin cellular decision making processes such as the regulation of cell division and end-cell function. Moreover, it is clear that the mechanisms by which cells achieve signal transduction specificity are the basis of developmental decisions.

The molecules that are responsible for the passage of intracellular signals can be thought of as the molecular switches of an intracellular integrated circuit, their discharge being tied to the receipt of an appropriate chemical signal. Extending the use of computer imagery is a useful, albeit limited, analogy for intracellular signaling, since in both cases there is a vectorial flow of information, a codified language by means of which the message is interpreted, (protein phosphorylation, the machine code of signal transduction!?) and a range of set responses by which the cell (computer) can "respond" to the inputs it receives. The limitations of the analogy become apparent when one looks a little deeper into the molecular underpinnings of signal transduction. Here there is tight coordination of all of the signal transduction pathways which are triggered by the many different extracellular stimuli which impact upon a cell at any given moment; intracellular signal transduction is clearly a highly networked process. While we are becoming more aware of the potential increases in computing power wrought by the networking of many smaller computers, the extra sophistication that an integrated and regulated network requires places intracellular signaling in an entirely different league. What, then, are the central themes of intracellular signal transduction?

There are a number of key elements which are central to our notions of the workings of intracellular signal transduction.

(1) A cell surface receptor is required to receive an extracellular stimulus and convert it into an intracellular signal.
(2) One or a limited number of thematic signal transduction pathways must be activated in order for the intracellular signal to be interpreted within the cell.
(3) Some mechanism by which the cell can interpret the signal needs to be in place.
(4) A means by which the cell can respond to that signal by making a functional or developmental decision must be available.
(5) A means of networking all of the decisions that occur in the course of a cell's lifetime needs to be in place in order that a coordinated response can be made to the vast range of signals which bombard a cell at any given moment.

Evolution arrived at a solution to this list of requirements around 500 million years ago. At this time the adoption of a multicellular (metazoan) phenotype came about, followed by the rapid expansion of organisms of this type. This rapid expansion of metazoan life forms is known as the "metazoan radiation" and may well have lasted a mere 20 million years. Whether the sophisti-

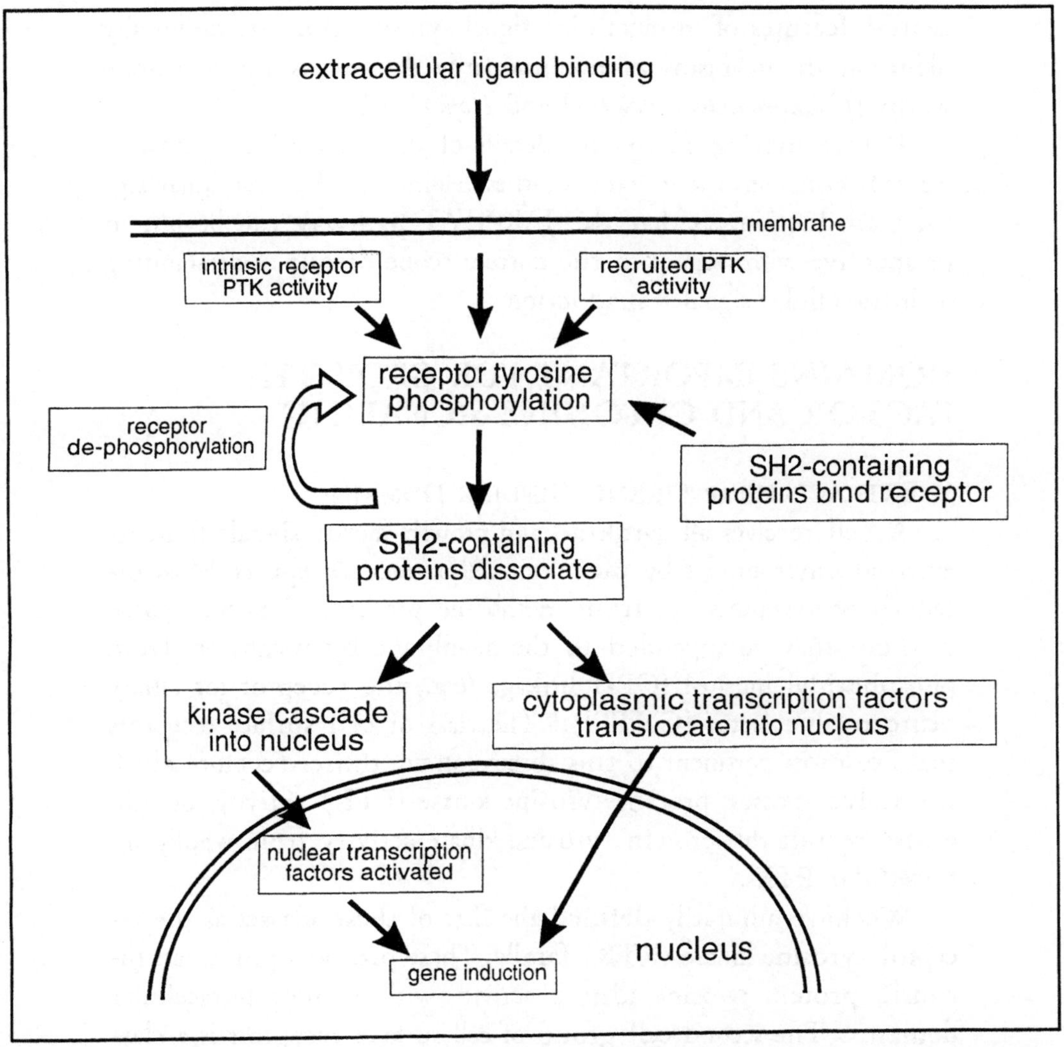

Fig. 1. Generalized scheme for intracellular signal transduction. Each of the key events in the cascade of signal transduction processes which follow extracellular binding of a ligand is outlined. Two pathways are shown, the ras/MAPK pathway and the JAK-STAT pathway.

cated signal transduction mechanisms we now recognize in the response of a cell to, say, epidermal growth factor (EGF), were in place in the protozoans which existed prior to the metazoan radiation, or whether the invention of one consolidating protein domain (such as the SH2 domain or the PTK domain) was the trigger, is not (nor ever can be) known. Importantly, however, the central features of intracellular signal transduction are essentially identical in organisms as evolutionarily distant as the nematode worm *(Caenorhabditis elegans)* and *Homo sapiens.*

Before moving on to the details of the JAK-STAT pathway, we will consider briefly each of these elements of intracellular signal transduction, so that the JAK-STAT pathway can be put in perspective with respect to the current scope of our understanding of intracellular signal transduction.

DOMAINS IMPORTANT FOR GROWTH FACTOR AND CYTOKINE SIGNALING

1. Extracellular Ligand Binding Domain

A cell receives all cytokine and growth factor signals from its external environment by means of cell surface receptors. Most often these receptors are trans-membrane proteins, however, some of them may be appended to the membrane by means of glycan phosphatidyl inositol (GPI) linkage (e.g., the receptor for ciliary neurotrophic factor [CNTF]).[1] The class of cell surface receptors that are most pertinent to this discussion are those receptors which themselves possess protein tyrosine kinase (PTK) activity, or otherwise recruit this protein tyrosine kinase activity from wholly intracellular PTKs.

We have arbitrarily defined the first of these classes as the receptor tyrosine kinase (RTK) family. These proteins possess an intrinsic protein tyrosine kinase activity within their intracellular domains. The second cell group of cell surface receptors is a class that we have arbitrarily called the *cytokine receptors.* These receptors are structurally similar to the RTK family of receptors, however, they do not possess within their intracellular domains protein tyrosine kinase activity. Nonetheless, it is now clear that there are common signal transduction mechanisms which lie downstream of each class of receptor, and that the general principles which underlie the activation of these receptors and the interpretation of the signals which are instigated by them, are by and large held in common.

2. The Protein Tyrosine Kinase Domain

Protein Kinases

Protein kinases are phosphotransferases capable of transferring the γ-phosphate from (usually) ATP to an acceptor amino acid on a substrate protein. These enzymes have been officially divided into four categories, based upon the particular acceptor amino acid recognized by each category. The two main classes are the protein serine/threonine kinases (STKs) and the protein tyrosine kinases (PTKs). The protein kinase domains of both classes of enzyme are approximately 250-300 amino acids in size. The catalytic domains of these enzymes are evolutionarily highly related, and both classes of enzyme presumably share a common ancestor.[2] Consequently, a significant degree of sequence identity emerges upon amino acid sequence comparison of the catalytic domains of members of these two classes of protein kinase.[3-5] Importantly, this sequence identity appears as a series of very highly conserved (often completely conserved) amino acid motifs embedded within regions of relatively poorly conserved amino acid sequence. Eleven of these highly conserved short amino acid motifs have been uncovered by careful analysis of a global sequence comparison of all members of the protein kinase family,[4,5] and each of these motifs have been assigned a roman numeral from I (being the most N-terminal motif), through to XI (the most C-terminal motif). A more thorough discussion of this aspect of the protein kinase family appears in other reviews.[4,6]

The generation of X-ray crystal structures of members of the STK and PTK families has been achieved, and this has provided the structural underpinnings for the notion that the aforementioned conserved motifs may play an important role in a number of aspects of the phosphotransferase activity of these protein kinase domains. Thus, it has been possible to deduce that subdomains I and II and VII play an important role in ATP binding,[7] while subdomain VIII plays a major role in the recognition of peptide substrates. In the tyrosine kinase family, an intriguing difference in the amino acid immediately prior to the key motif in subdomain VIII, -KWTAPE- or -KWMAPE- seems to be the basis for a degree of substrate specificity in this branch of the family of kinases.[8]

Protein Tyrosine Kinases

The subgroup of the protein kinase family known as the protein tyrosine kinases (PTKs) are a thematic protein family (comprising at least ten structural variants),[7] built upon a variant of the kinase domain which appears to confer the recognition of tyrosine for substrate specificity. The physiologically powerful catalytic activity of this domain has been harnessed to serve a wide range of physiological demands including functions as diverse as transduction of extracellular growth and differentiation stimuli (reviewed in ref. 9) and cellular response to changes in intracellular REDOX potential.[10] Members of this family are composed of a mosaic of sensory, regulatory and effector domains, an organization best understood in one structural variant of the family: the growth factor receptor PTKs (receptor tyrosine kinases or RTKs). The structure of signal transduction molecules such as these facilitates a unidirectional flow of information from the outside of the cell to the inside of the cell. This process is specific for the correct receptor/ligand interaction, and results in significant amplification (or "gain") as the signal (phosphorylation of substrates upon tyrosine residues) is transmitted within the cell. Under the strictest control, this combination of specificity and thematic signal transduction has been frequently used in building the repertoire of possible differentiation processes which are available to the totipotent cells of a pre-implantation embryo. Unconstrained, it is also, by virtue of its pivotal position in most widely accepted schemes of signal transduction processes, a likely mechanism by which cancer cells acquire some of the features of their tumorigenic phenotype, such as their apparent independence of extracellular growth stimulation (reviewed in refs. 11, 12).

The PTK family, with approximately 70 known members within the mammalian genome, is somewhat smaller than the STK family, with around 200 known mammalian family members (Runting AS and Wilks AF, unpublished work). The discovery of PTKs from organisms belonging to other metazoan phyla has swelled the number of PTKs to close to the 100 mark, and has helped to define the notion that this class of protein has an important role to play in the intracellular signal transduction processes which underpin the processes of cellular differentiation and

Table 1. The family of cytoplasmic PTKs

Family	Size	Key Features	Family Members
The *SRC* Family	55-60 kD	SH2 domain SH3 domain	SRC,YES, FYN, FGR, LYN, HCK, LCK, BLK.
Tec Family	60-73 kD	SH2 domain SH3 domain Plextrin Homology domain	Tec, Emt, Btk.
csk Family	50 kD	SH2 domain SH3 domain	csk, HYL
fes Family	93-96 kD	SH2 domain	Fes, Fer
Abl Family	150 kD	SH2 domain SH3 domain Nuclear Localisation Sequence DNA binding domain?	Abl, Arg
Syk/ZAP-70 Family	70-72 kD	2 X SH2 domains	Syk, ZAP-70
Fak	120-125 kD	Focal Adhesion Targetting domain **No** SH2 or SH3	Fak
JAK Family	130 kD	Kinase-like domain **No** SH2 or SH3	Tyk2, JAK1, JAK2, JAK3

An approximate molecular size for a representative of each family of PTKs is shown alongside a list of the functional domains that they possess. Other mammalian members of each family are listed in the left-hand column of the table.

development. Indeed, the absence of protein kinases which are identifiably PTKs (based upon either sequence comparison with other members of the known PTK family, or on the basis of functional tests which might measure amino acid acceptor specificity, for example) from protozoan eukaryotes, suggests that PTKs may well be a defining feature of the metazoan "lifestyle." Indeed, their development could even be one of the causative events which helped eukaryotes develop a multicellular organization during the 10 million years of evolution which define the metazoan radiation (see above).

Nonreceptor protein tyrosine kinases

There are nine subfamilies of nonreceptor PTK; these are listed in Table 1 and shown in Figure 2. Each structural variant presumably has been tailored to act in a particular metabolic niche within the cell. While in most cases the precise cellular role of each of these classes is not known, the presence of particular

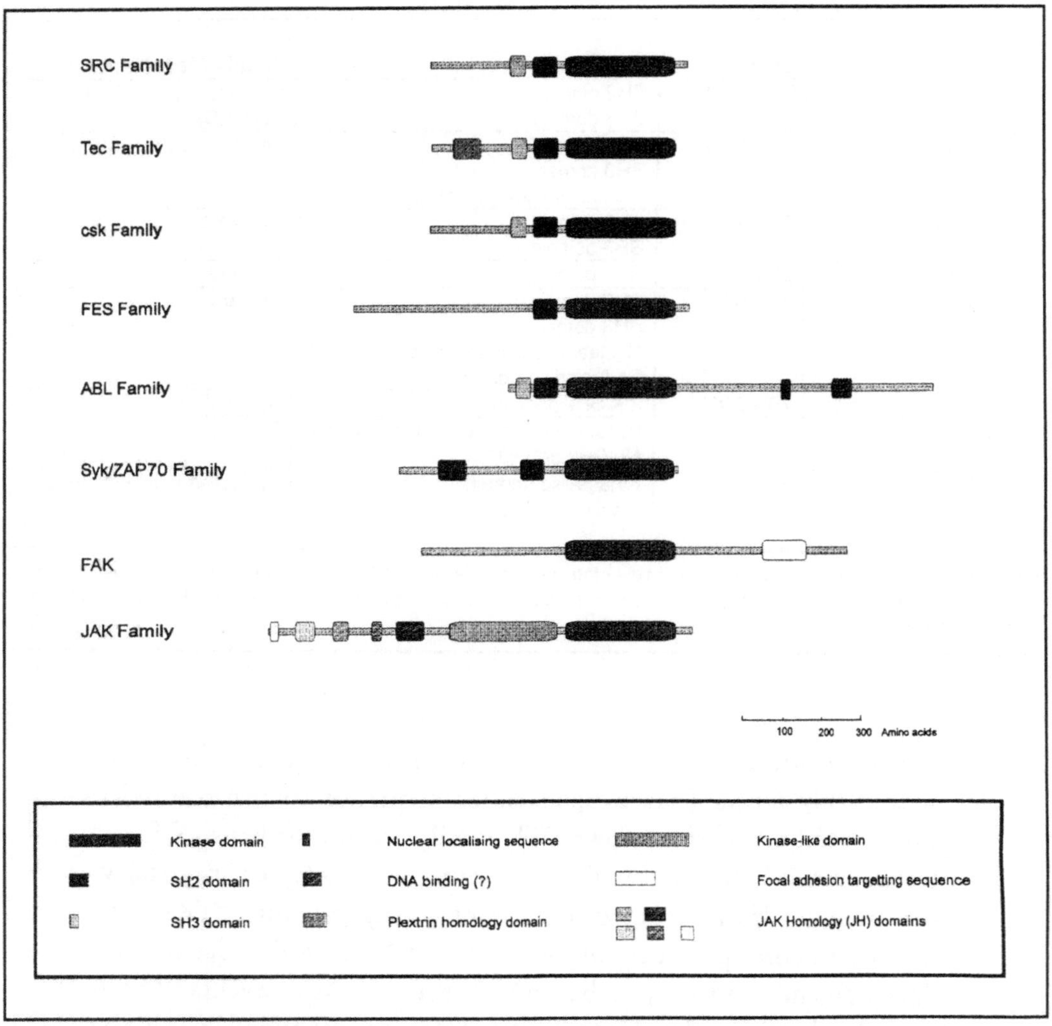

Fig. 2. Cytoplasmic protein tyrosine kinases. A representative of each family of intracellular PTKs is shown to scale. The PTK domain is the feature held in common by each of these proteins. Extracatalytic domains (such as SH2, SH3 and plextrin homology domains) important for the function of these proteins are also shown, with their approximate location on the molecule.

conserved domains, such as SH2, SH3 and plextrin homology domains, may ultimately give clues as to what their cellular role might be.

Receptor protein tyrosine kinases

The receptor protein tyrosine kinases (RTKs) are a large family of molecules which, in addition to possessing a PTK domain, also have an extracellular ligand binding domain. This places this class of PTKs at the cell membrane, and as a consequence these proteins are of pivotal importance for the transmission of developmental cues during embryogenesis, a notion underlined by the fact that many developmental mutants in both mouse and *Drosophila melanogaster* are allelic to genes encoding PTKs.

As a general rule the intracellular domains of RTKs possess one or more specific tyrosine residues which become phosphorylated in response to ligand binding to the RTK extracellular domain. These tyrosines are often located in the region between the C-terminal end of the PTK domain and the C-terminal tail of the protein, as is the case for the EGF-R. Several classes of PTK possess an additional protein domain, inserted between the two lobes of the PTK domain, and it is a common feature in this arrangement that a number of substrate tyrosine residues are located in this domain. The Class III RTKs (which include the platelet-derived grown factor [PDGF-R] family) are good examples of this sort of arrangement. It is now widely accepted that these tyrosine autophosphorylation sites are SH2 domain binding sites (see below and refs. 13, 14). Thus, the PDGF receptor autophosphorylation sites are binding sites for the SH2 domains of phospholipase C-γ1, GTPase activating protein (GAP), PI3'-Kinase and *SRC* tyrosine kinase,[15,16] while the SH2 domain located on the p85 subunit of PI3'-Kinase recognizes phosphotyrosines on the EGF-R, colony stimulating factor (CSF1-R) and *c-kit*.[17] The recruitment of second wave signal transduction molecules is dealt with in more depth below.

3. The SH2 Domain

The SRC homology 2 (SH2) domain is a non-catalytic protein module of approximately 100 amino acids.[18,19] SH2 domains were identified initially as domains of conservation which were held

Table 2. Phosphotyrosine containing SH2-binding sites

SH 2 Domain	Sequence Specificity
SRC family	-YEEI- -.DNV- -.TDM- -...L-
Abl family	-YENP- -.TEV- -.MDL-
Sem 5 family	-YLNV- -.V.P- -.I..- -.M..-
SH-PTP family (N-terminal)	-YIXV- -.V.I- -...L- -...P-
STAT1	-YIKTE- -.KHRL- -.L.TK- -.D.PH-
vav	-YMEP- -.L..- -.E..-

The amino acid specificity of the SH2 domain of a representative of each family of molecules is shown. Where the SH2-recognition sequence varies the alternate amino acid is shown; dots (.) indicate identity between all recognition sequences for a given SH2 domain.

in common between a number of cytoplasmic PTKs, including the virally encoded oncogene *v-fps/fes*[18] and *v-src.*[19] Although this domain appeared to possess no intrinsic catalytic activity, it soon became obvious that the subdomain was of tremendous importance in the signal transduction processes which lie downstream of the activated oncogenic PTKs. Mutations within this region of the p130$^{gag\text{-}fps}$ transforming protein encoded by the Fujinami sarcoma virus (FSV), suppressed the cellular transformation capacity which was normally induced by the wild type oncogene.[18,20] Thus PTK signal transduction requires both a functionally active PTK domain and a functional SH2 domain.

Families of SH2-containing proteins

Since their discovery, SH2 domains have been observed in a large number of important signaling proteins. Essentially two classes of SH2-containing signaling proteins have been described (Fig. 3). Type 1 includes those SH2 domain-containing signaling proteins which also possess an intrinsic enzymatic activity. Such activities include guanine-nucleotide exchange (for example the *vav* protooncogene[21,22]) or protein tyrosine kinase activity (e.g., *src, fes, abl,* and *Syk*[7] inter alia). The second class of SH2 containing signal proteins includes those molecules which, on structural grounds appear to lack any intrinsic catalytic activity and are likely therefore to be adaptors. This class of molecules includes proteins such as NCK,[23] GRB2[24,25], p85 subunit of PI 3'-kinase[26] and the SHC protein. These adaptor molecules serve to act as bridging links between other functional components of a particular signal transduction pathway.

Function of SH2 domains

The function of the SH2 domain is the specific binding of phosphotyrosine-containing motifs. Thus, SH2 domains bind to the phosphotyrosine residues present on a variety of intracellular signal transduction proteins, such as those that are generated by virtue of the activation of the intrinsic protein tyrosine kinase activity of RTKs. The binding of SH2 domains to phosphotyrosine occurs with high affinity, and with a high degree of sequence specificity with respect to those amino acids which lie immediately N-terminal and immediately C-terminal to the phosphotyrosine itself.[13,27] The coordination of intracellular signal transduction requires the generation of specificity and selectivity if a particular extracellular signal is to be directed toward the generation of an appropriate physiological response. As already discussed, the activation of the PTK domains of RTKs results in the tyrosine phosphorylation of the intracellular domain of these RTKs generating a series of molecular tags (phosphotyrosine residues), to which proteins bearing SH2 domains can bind.

The binding of an SH2 domain to these molecular tags is achieved with both high affinity and specificity. Cantley and coworkers[13,14,28,29] have demonstrated that the specificity of the SH2

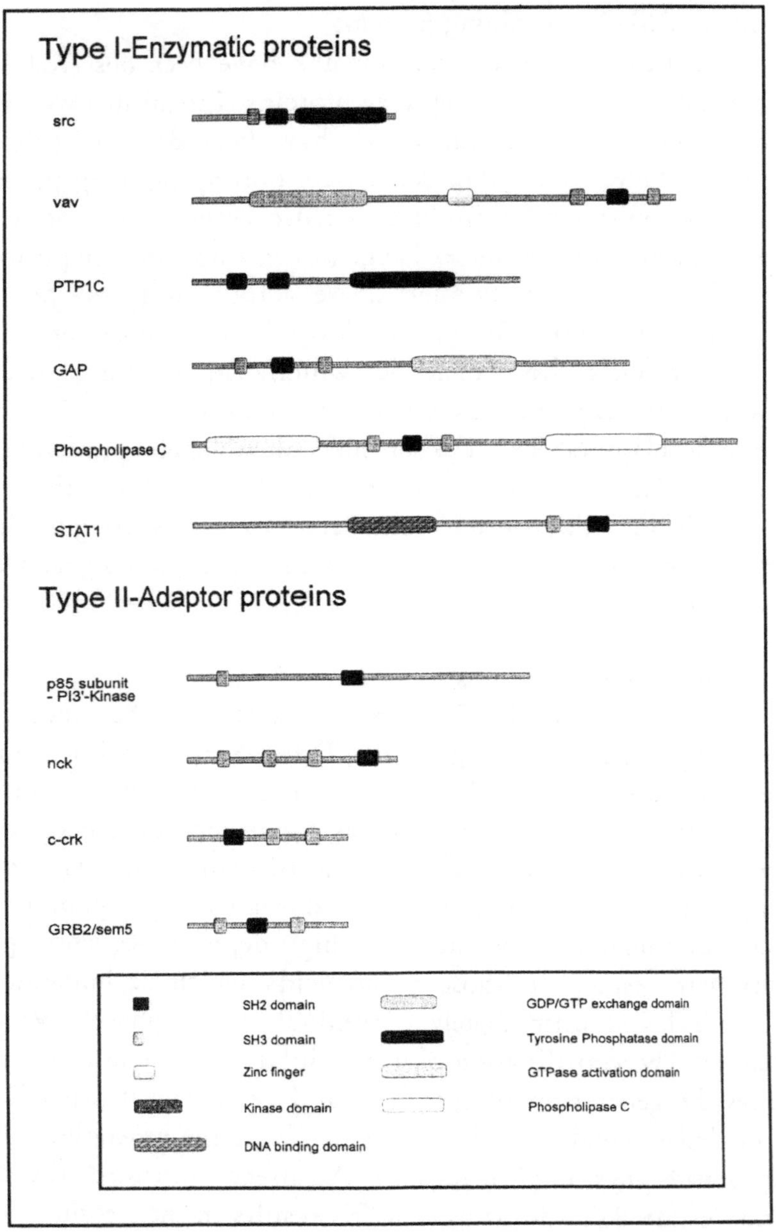

Fig. 3. SH2 domain-containing proteins. SH2 domain-containing proteins are divided into two broad categories: Type I proteins that couple an adaptor function to intrinsic catalytic activities, such as PTK activity or GTPase activation; and Type II proteins that appear to possess no intrinsic catalytic activity, most probably functioning as "adaptor" molecules. This class of molecule provides links with a range other components of signal transduction pathways. This figure was constructed along the lines of comparisons discussed elsewhere (see text for citations).

domains of a wide range of signal transduction molecules is due to the recognition of amino acids surrounding the phosphotyrosine residue. A list of SH2 domains and their amino acid recognition sequence specificity is included in Table 2. A systematic approach to the identification of the sequence specificity for high affinity interaction with SH2 domains by phosphotyrosine containing peptides was undertaken using degenerate phosphopeptide libraries.[13,14] From these studies it was clear that residues lying within the four amino acids C-terminal of the phosphotyrosine residue are of particular importance to substrate specificity.

Structure of SH2 domains

The structural basis for this binding specificity was elucidated following the determination by X-ray crystallography of the three-dimensional structure of SH2 domains from a diverse range of signal transduction molecules including c-*abl*,[30,31] p56 *lck*[32,33-35] and others. Other studies undertaken using multidimensional nuclear magnetic resonance (NMR) spectroscopy have been undertaken for a number of SH2 domains. Taken together, the X-ray crystallography and NMR spectroscopy structures are coherent and demonstrate that the SH2 domain is a compact and largely spherical entity with an hydrophobic core enclosed by a pair of three-stranded antiparallel β sheets and a C-terminal α helix. All SH2 domains for which the structure has been determined conform largely to this basic model. The cocrystallization of the SH2 domain of v-*src* with two different phosphopeptides has been also undertaken[35] and a cocrystallization of the SH2 domain of c-*abl* with a substrate peptide has also been achieved.

These studies have generated important information regarding the precise molecular interactions which define the sequence specificity of phosphotyrosine/SH2 interactions which are occurring during the process of signal transduction. The N- and C-termini of the SH2 domain meet at the base of the module and the phosphotyrosine binding site is exposed at the opposite surface. Two pockets are formed on this binding surface, one of which is generated by the highly conserved motif (FLVRES) and this is the site of docking for the phosphotyrosine residue itself, the arginine of the FLVRES motif being involved in the interaction with the phosphate group from the phosphotyrosine. The second binding pocket is required for the interaction with the amino acid which is in the +3 or +4 position C-terminal to the phosphotyrosine residue, and is therefore required to generate the specificity of interaction of the phosphopeptide with the SH2 domain.

While a considerable degree of specificity can be generated by locating particular amino acids in the +3 hydrophobic pocket, additional specificity appears to be generated by means of the interaction of SH3 and SH2 domains. For example, studies examining the affects of SH2 and SH3 domains on the regulation of the catalytic activity of c-*SRC* revealed that c-*SRC* mutated specifically within the SH3 domain was no longer able to assume a conformation that interacted with the C-terminal phosphotyrosine containing polypeptide tail. This led to the speculation that the SH3 domain was playing a role in the interaction of the SH2 domain with this phosphotyrosine residue.[36]

Recent crystallographic studies of substrate peptide interactions with the *lck* SH2 and SH3 domains have now confirmed the likely interaction of these two domains.[33] The crystal structure of a portion of the *lck* SH2 and SH3 domains, both with and without phosphopeptide, revealed a dimeric structure involving intermolecular interactions between the SH2 and SH3 domains. In this molecule, the two domains are separated by only two amino acids and the interaction involves an hydrophobic "pouch" in the SH3 domain, which interacts with residues in the SH2 domain in a manner similar to that seen for the proline rich motifs,[37] with the phosphorylated tail folding back to interact with both the SH2 and SH3 domains. While this close interaction between SH2 domains and SH3 domains in members of the *SRC* family of PTKs might account for the subtleties of phosphopeptide sequence recognition within this family, it does not yet offer any functional significance for such a colocalization of SH2 and SH3 domains within adaptor molecules such as SHC.

In summary, the PTK/SH2 domain nexus is clearly pivotal to the generation of signal transduction specificity in eukaryotic cells. This specificity appears to be generated in two ways, firstly by selection of phosphotyrosine substrates by the PTK domain, and secondly, by the selective binding of particular phosphotyrosine residues by SH2 domains. This dual level of specificity largely accounts for the specific signal transduction pathways which are activated in response to the extracellular binding of a ligand and a surface receptor. This PTK/SH2 domain nexus is of crucial importance to the JAK-STAT pathway, and is outlined in subsequent chapters.

THE "CASCADE" MODEL OF SIGNAL TRANSDUCTION

Intracellular events arising from RTK activation include protein phosphorylation, inositol phospholipid turnover, Ca^{2+} influx, Na^+/H^+ exchange and the activation of gene expression.[9,38,39] For the purposes of this brief review, however, the focus will primarily be on the downstream protein phosphorylation events, and in particular, those pathways utilizing SH2-containing proteins.

Amongst the most extensively characterized of the RTKs are the receptors for epidermal growth factor (EGF-R) and platelet-derived growth factor (PDGF-R). Each of these receptors has been shown to bind a wide range of intracellular signaling molecules, and they therefore provide useful paradigms for exemplifying the general features of signal transduction. Signal transduction downstream of the RTKs can be defined as the "Cascade" pathway of signal transduction, and contrasts with what might be called the "Direct-Effector" or JAK-STAT model (Fig. 1).

We have broken down the process of signal transduction into its component parts and will briefly examine each of them in order that the general principles of receptor activation and signal transduction may be understood for future chapters.

1. Receptor Activation

The general mechanism of RTK and cytokine receptor activation is a process of ligand mediated dimerization. This process can be precipitated by one of at least three distinct modes.

(1) Dimeric growth factors such as PDGF may induce dimerization of their cognate receptors by bridging two distinct PDGF receptor chains.[40]

(2) A monomeric ligand possessing at least two distinct receptor binding sites may be capable of inducing dimerization by means of interaction with each of these

binding sites on distinct receptor chains. A good example of this is the growth hormone receptor (GHR) which is dimerized by binding to a single growth hormone molecule.[41]

(3) The third mechanism is where receptor dimers pre-exist as membrane complexes prior to the interaction of ligand with receptor. One example of this format is the insulin receptor[42] which exists as a homodimerized molecule held together by intermolecular disulphide linkages. This process of ligand mediated dimerization of the extracellular domains of growth factor receptors is the key to the induction of the activities held in the intracellular domains of these proteins.

2. "Second Messenger" Signal Transduction Molecules

Different SH2-containing molecules may bind to the same activated receptor at distinct sites, while an individual SH2-containing molecule may be activated in response to a number of different growth factors or cytokines. In this way permutations of a limited number of signaling molecules may be brought to bear to define a wide range of cellular responses, each response being tailored to the particular growth factor, and to the cell type in which the response is taking place. A generalized scheme for the ras-MAPK pathway is outlined in Figure 4. Details of the individual components of this pathway are presented below.

GRB2/Sos complex

GRB2 is a 23 kDa adaptor protein consisting of an SH2 domain flanked by two SH3 domains. This adaptor binds to phosphotyrosine residues on the EGF-R,[43,44] insulin receptor (IR),[45] insulin receptor substrate-1 (IRS-1) signaling molecule,[46,47] and others.[24,48] Upon EGF stimulation, GRB2 binds to the EGF-R via its SH2 domain, and then in turn to the nucleotide exchange factor, son of sevenless (Sos), via its SH3 domain. The Sos/GRB2 complex is then recruited to the plasma membrane as a consequence of the interaction with the membrane-associated *ras* guanosine triphosphatase (GTPase).[44] As the intrinsic activity of Sos does not increase after EGF stimulation, it appears that this

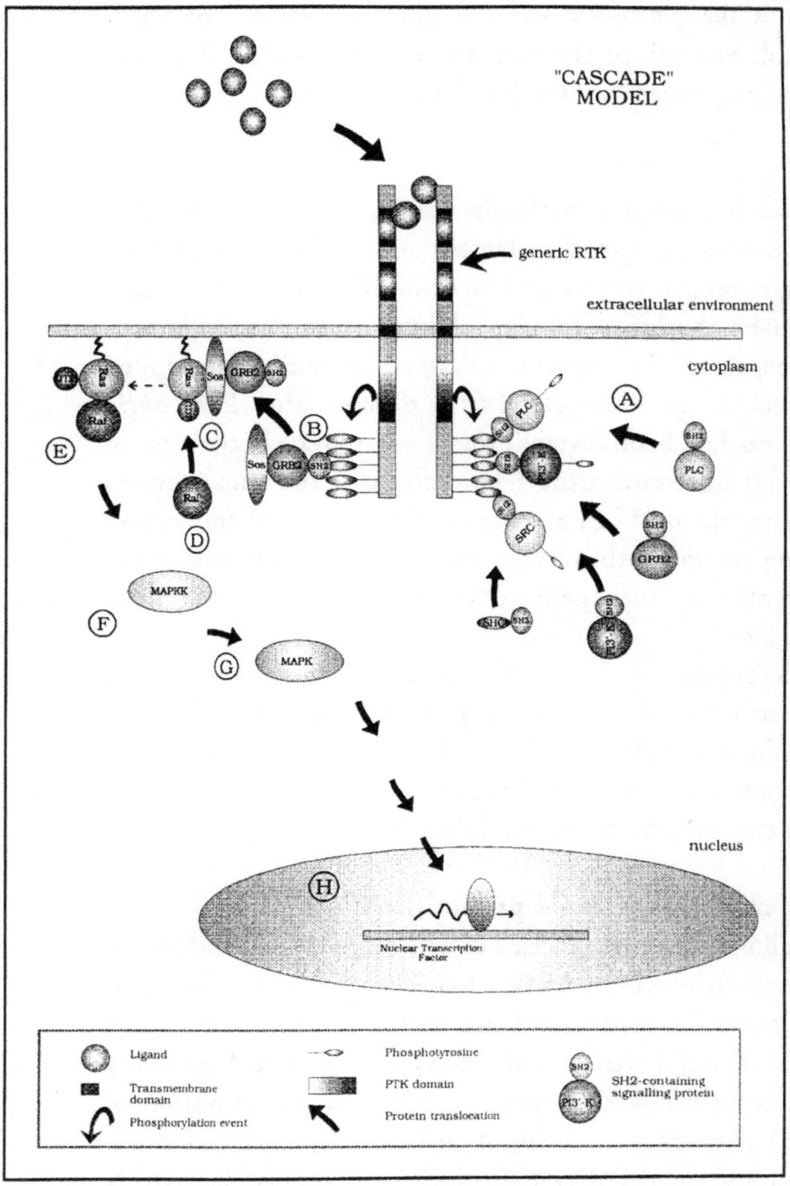

Fig. 4. The "Cascade" or ras/MAPK signaling pathway. This figure depicts events described in detail in the text. The size of individual signal transducing molecules is not to scale with respect to relative cellular components. Ligand binding results in receptor dimerization and activation. Receptor activation leads to auto- and trans-phosphorylation of the receptor intracellular tail via the intrinsic catalytic domains of the receptor chains, which in turn provide binding sites (A, B) for intracellular signaling molecules such as GRB2 and PI3'-K. These receptor-bound SH2-containing molecules are then activated and move off the receptor (C). In the case of GRB2/Sos, this results in the activation of the MAP Kinase pathway via ras and raf (D, E, F). This leads to the phosphorylation and activation of MAP Kinase (G), which then moves to the nucleus where it activates nuclear transcription factors to initiate gene induction (H).

translocation is the necessary event in the "activation" of Sos.[44] Sos is then able to catalyze the exchange of GDP for GTP on *ras*, thereby generating active *ras* (reviewed in ref. 49).

ras GTPase

ras GTPase is a pivotal molecular switch for signal transduction pathways between tyrosine kinases and serine/threonine kinases, leading to cellular differentiation or proliferation. This critical role is shown by the large number of human tumors with activated *ras* oncogenes.[50] In concert with genetic evidence compiled from experiments in *Caenorhabditis elegans* and *Drosophila melanogaster*, *ras* has been established as a universal regulator of cell growth and differentiation.[51] Furthermore, the link between tyrosine kinases and *ras*-GTPase was confirmed by microinjection of neutralizing *ras* antibodies. Such microinjection prevents cellular transformation by oncogenic tyrosine kinase but not serine/threonine kinases.[52]

ras is a GTP-binding protein that cycles between the inactive, GDP-bound state and the active, GTP-bound state, and there are many extracellular signals stimulating the formation of the active *ras*-GTP.[53] A major downstream function of *ras*-GTP is to control the MAP kinase cascade, described below.[54]

Activators of mitogen activated protein (MAP) kinase

The signaling molecules utilized downstream of GRB2/Sos/*ras*, specifically those involved in MAP kinase activation, are also common to a number of receptor pathways (reviewed in ref. 49). One such component, Raf-1 (discussed below) is modulated by many distinct pathways (reviewed in refs. 54, 55). Briefly, activation of *ras* (*ras*-GTP) triggers a cascade of serine/threonine and dual-specificity kinases, the MAP kinases and their regulators. MAPK is the last step in this cascade. *ras*-GTP binds to the serine-threonine kinase Raf-1 (also known as MAP kinase kinase kinase [MAPKKK]) which then phosphorylates MAP kinase kinase (MAPKK). MAPKK then activates MAP kinase.

Raf-1

Raf-1, a serine/threonine-specific protein kinase, is one of a number of cellular kinases known to activate MAP kinase kinase (MAPKK). Active *ras* (*ras*-GTP) binds the N-terminal domain of Raf-1.[56-60] After binding to Ras-GTP and becoming tyrosine phosphorylated, Raf-1 stimulates MAPKK. For example, following IL-2 stimulation, Raf-1 becomes tyrosine phosphorylated by the activated *SRC* kinase ($pp60^{v\text{-}SRC}$).[61,62] The tyrosine phosphorylation of Raf-1 is absolutely required in conjunction with binding to Ras-GTP to activate the dual specificity kinase, MAPKK.[61] Raf-1 is fully activated by a number of factors, such as protein kinase C (PKC), *ras*-GTP and activated *SRC* kinases.[54] However, these factors may not all lead to the same final result, rather, a variety of responses may emerge. For instance, Raf-1 is phosphorylated by PKC, following phorbol ester treatment, but this leads to increased autophosphorylation of Raf-1 and not to the activation of MAPKK.[54] Once activated by Raf-1, MAPKK is primed to target the MAP kinase isoforms.

MAP kinase and MAP kinase kinase

The activation and nuclear translocation of the cytoplasmic serine/threonine MAP kinase isoforms (Erk-1 and Erk-2) is the net result of upstream *ras* activation.[55] MAP kinase (MAPK) is activated by phosphorylation on threonine and tyrosine by MAPKK (after it is activated by Raf-1, as described above). MAPK then phosphorylates and thus, activates nuclear transcription factors including c-MYC, c-JUN, c-Fos, nuclear factor-interleukin-6 (NF-IL-6), the cell surface proteins cytoplasmic phospholipase A_2 ($cPLA_2$), EGF-R and protein kinases such as c-Raf-1, MAPKK and $p90^{rsk}$ (the 160 kDa glycogen binding subunit of protein phosphatase-1 [PP-1]; reviewed in ref. 55). This pathway is a direct example of the compartmentalization of various signal transduction events, whereby transcription factors reside within the nucleus and the activating kinases translocate to this compartment.

Other PTK-activated second messengers

There exist a number of SH2-containing proteins that bind activated transmembrane receptors that are independent of the *ras* pathway. PI3'-K and PLCγ are two such proteins, and these will be introduced to further reinforce the notion that the SH2-phosphotyrosine nexus is an extremely versatile, and intrinsically specific mechanism for directing transduction pathways.

Phosphatidylinositol-3' kinase (PI3'-kinase)

PI3'-K is an enzymatic activity comprised of a catalytic subunit (p110), and an adaptor subunit (p85). This enzyme phosphorylates the D-3 position on phosphatidylinositol, phosphatidylinositol-4-phosphate [PI(4)P], or phosphatidylinositol-4,5-bisphosphate [$PI(4,5)P_2$] to produce PI(3)P, $PI(3,4)P_2$, and $PI(3,4,5)P_3$[38,49] respectively. The p85 subunit has an amino-terminal SH3 domain, a central Rho-GAP homology domain of as yet undefined function, and two C-terminal SH2 domains. The "Kinase Insert" phosphotyrosine residues Tyr740 and 751, on the PDGF-R have been demonstrated to be high affinity binding sites for the SH2 domain of PI3'-K. In addition, there are a number of other known phosphotyrosine binding sites for the SH2 domains of the p85 subunit such as those present on activated IRS-1, TCR-associated CD28 and CSF1-R.[63-66] The SH2 domain on the adaptor subunit, p85, binds to an activated receptor, whereby it is capable of recruiting, via its SH3 domain, the p110 catalytic unit.

The p110 catalytic subunit exhibits homology to the yeast protein VPS34, known to sort proteins to vacuoles. While a function for p110 is yet to be formally defined, there are suggestions of a similar activity to VPS34. This is supported by the observation that mutant PDGF-Rs that are unable to associate with PI3'-K, fail to become localized to the juxtanuclear region of the cell following PDGF stimulation (reviewed in ref. 49).

While the cellular function of PI3'-K remains elusive, a role in mitogenic signal relay looks likely. The activation of the PDGF-R leading to a stable association with PI3'-K results in the stimulation of the serine/threonine kinase, $p70^{S6K}$.[49] $p70^{S6K}$ appears to be required for serum-induced protein synthesis, c-Fos induction and entry into cell-cycle S phase.[67]

Phospholipase C-γ

The phospholipase C-γ isoform, PLC-γ1, is a protein capable of cleavage of $PI(4,5)P_2$ to inositol triphosphate (IP_3) and diacylglycerol (DAG), after activation by phosphorylation on tyrosine 783.[49] Both IP_3 and DAG are critical second messengers, mediating the release of Ca^{2+} from intracellular stores, and the activation of PKC, respectively.[68,69] The release of Ca^{2+} does not appear to be required for a mitogenic response, and therefore it is considered that it is the activation of PKC that leads to mitosis. This is supported by the observations that potent tumor promoting phorbol esters activate PKC. In addition, the overexpression of diacylglycerol-activated PKC isoforms lead to deregulated cell growth and cellular transformation.[49]

PLC-γ1 possesses two SH2 domains, which have been shown to interact with the phosphorylated tyrosine 992 located in the C-terminal tail of the activated EGF-R,[70] and phosphotyrosine 766 on the EGF-R.[71] In crystal structure studies, the C-terminal SH2 domain has been shown to bind to a phosphopeptide corresponding to the tyrosine 1021 of the PDGF-R.[72,73]

This interactive molecule links growth factor RTKs to the so-called "second messenger" pathways of the inositol phospholipids, and is thus a further example of the degree of overlap between pathways, both in terms of the signaling molecules and the mechanisms they utilize. SH2 domains have been shown to be vital components in different types of intracellular signaling pathways, including the *ras*-dependent and inositol phospholipid pathways.

Insulin receptor substrate-1

It is apparent now that not all RTKs actually directly mediate the association of all the SH2-containing signaling proteins that they utilize. The receptors for insulin and insulin-like growth factor (IGF-1) both possess intrinsic catalytic domains, the substrate of which is both itself and the insulin-receptor substrate 1 (IRS-1). This protein is almost the equivalent of an "off shore" kinase insert domain, and possesses docking sites for many of the known SH2-containing signaling molecules described above (including GRB2, p85 subunit of PI3'-K and SH-PTP2; reviewed in ref. 74).

IRS-1 is a large phosphoprotein (~180 kDa) that was first identified from antiphosphotyrosine immunoprecipitates following insulin stimulation.[75] It has a high phosphoserine content that results on SDS page in the protein migrating anywhere between 165-185 kDa.[76,77] There are 20 potential tyrosine phosphorylation sites, and a number of these sites have been demonstrated to be substrates for the PTK domain of the insulin receptor, and the IL-4R-associated PTK, as well as docking points for SH2-containing signaling proteins.[77]

The insulin and IGF-1 receptors are not alone in their utilization of IRS-1. The IL-4 receptor, while having to recruit cytoplasmic tyrosine kinases such as JAK1, or Fyn, is capable of inducing the tyrosine phosphorylation of IRS-1 also. Thus, an intrinsic catalytic domain appears not to be a prerequisite for the involvement of IRS-1. A similar theme appears to be in place within hemopoietic cells, whereby a protein substrate called 4PS that is functionally related to IRS-1 is phosphorylated in response to insulin, IGF-1 and IL-4, and similarly associates with PI3'-K.[78]

DISCUSSION

We have presented herein the key elements of the "Cascade" or ras-MAPK signaling pathway. Two key notions emerge from this pathway. Firstly, the paramount importance of the phosphotyrosine/SH2 domain nexus to intracellular signaling. Secondly, the ras/MAPK pathway is both complex and intricate. This pathway is clearly an important feature of the intracellular response to the triggering of both RTKs and cytokine receptors. Could the ras/MAPK pathway fulfill all of the requirements of such a diverse range of signaling molecules? In the coming chapters we will examine an alternative signaling pathway which is also held in common by both RTKs and cytokine receptors, but one that makes a much more direct approach to the nucleus in order to regulate gene transcription. Herewith the second major SH2/phosphotyrosine mediated signal transduction pathway; the JAK-STAT pathway.

CHAPTER 2

The Cytokine Receptor Family

The cytokine receptors comprise an evolutionarily related group of molecules which fall into two distinct classes; namely the class I and class II cytokine receptors[79-81] (Figs. 5-9). Class I comprises the cellular receptors for many interleukins (IL), such as those that regulate hematopoiesis, e.g., IL-3,[82] IL-5, IL-11, granulocyte-macrophage colony stimulating factor (GM-CSF),[83,84] granulocyte-colony stimulating factor (G-CSF),[85,86] erythropoietin (EPO),[85,87] and thrombopoietin (TPO),[88,89] those that regulate lymphocyte production and function, e.g., IL-2,[90-92] IL-4,[93] IL-9,[94] IL-7, IL-12, IL-13, IL-14 and IL-15, and those that have more pleiotropic effects such as growth hormone (GH), prolactin (PRL), IL-6,[95] leukemia inhibitory factor (LIF),[96] ciliary neurotrophic factor (CNTF)[97] and Oncostatin M (OSM). The class II cytokine receptors include the receptors for interferon γ (IFNγ),[98] interferon α (IFNα)[99,100] and IL-10.[101]

GENERAL FEATURES OF CYTOKINE RECEPTORS

Hemopoietin Receptor Domains and FNIII Repeats

The hemopoietin receptor domain (HR) is composed of two subdomains of approximately 100 amino acids, the most N-terminal of which characteristically possesses four conserved cysteine residues, while the C$^-$ terminal region contains a conserved tryptophan-rich sequence element (WSXWS).[79] The recent determination

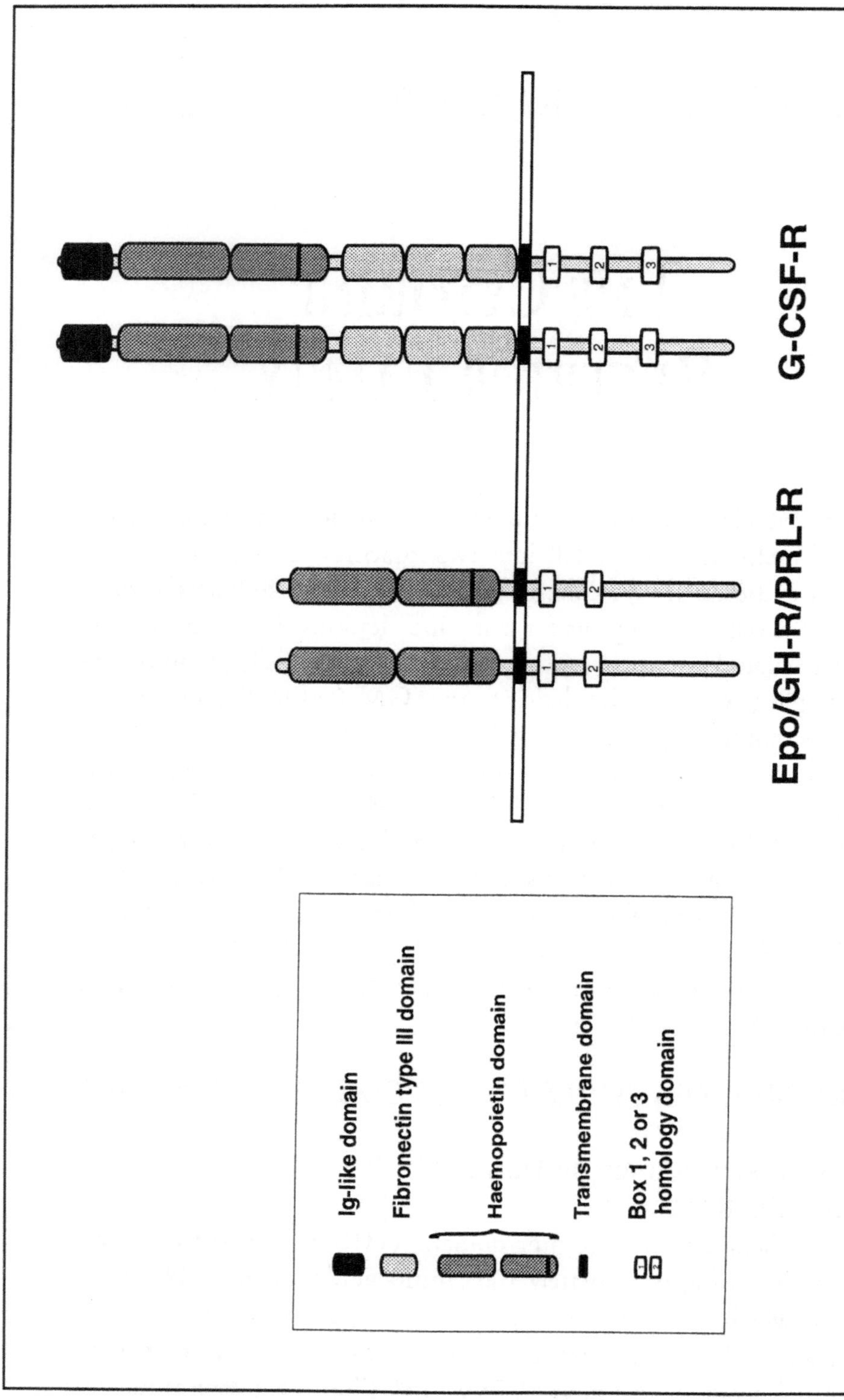

Fig. 5. Schematic of the arrangement of domain structure of homodimerizing cytokine receptors. The domain structures of the EPO/GH/PRL receptors are shown, alongside the structure of the G-CSF receptor. The EPO receptor is composed of a single "hemopoietin domain", whereas the G-CSF receptor in addition bears three fibrinectin type III repeats and a single Ig-like domain.

of the X-ray crystallographic structure of the growth hormone receptor (GHR) showed that the two hemopoietin subdomains are held at a 90° angle to each other.[102] Each of these domains consist of seven β-strands that form a sandwich of two antiparallel β-sheets of four and three strands.[102] The overall structure of the hemopoietin domain most resembles that of the D2 Ig-domain of CD4, an atypical Ig super-family member. In addition to the possession of a hemopoietin domain, many of the cytokine receptors possess a fibronectin type III (FNIII) domain. This is an approximately 90 amino acid motif which is possessed in varying numbers by several members of the cytokine receptor family.

The basic hemopoietin domain is found either alone (e.g., EPO-R GH-R etc.), in combination with other domains such as Ig-like domains (e.g., IL-5Rα, IL-3Rα, GM-CSF-Rα) or in combination with Ig-like domains and FNIII domains (e.g., gp130, LIF-R, G-CSF-R) The most unusual format is that exhibited by the LIF-R which possesses in order of appearance from the transmembrane domain to the N-terminal, three fibronectin type III (FNIII) domains, a hemopoietin domain, an Ig-like domain, and a second duplicated HR domain. The significance of this baroque architecture is unknown.

The possession of at least two sets of common sequence elements in the class I and class II receptors suggests a degree of evolutionary relatedness. This is not surprising considering the similarity of many of the ligands for these receptors, in that they themselves are members of the "four helix bundle" family of molecules. Thus, both the ligands and the receptors of this system appear to have co-evolved to undertake a host of physiological functions.

Box 1 and Box 2

Several conserved sequence motifs have been identified within the intracellular domains of the cytokine receptors. These motifs have, in turn, been demonstrated to be important for the recruitment of downstream signaling molecules. For example, two elements (referred to as "Box 1" and "Box 2") have been uncovered in the cytoplasmic domains of most class I cytokine receptors. It is now apparent that these domains are crucial for the transmission of signals downstream of the receptor. Box 1 is a proline-rich element which lies in the membrane proximal region of the intracellular domain of all of the class I cytokine receptors[103] whereas Box 2 is a relatively acidic element located some 40 amino acids C-terminal to the Box 1 element.[103] Deletion mutation analysis of a number of class I cytokine receptors, (e.g., GH-R,[104,105] and G-CSF-R[103,106]) has underscored the importance of each of these domains to the process of intracellular signaling. For the GH-R,[104,105] analysis of a series of deletion mutants generated by truncating the C-terminal end of the intracellular domain, revealed that the majority of the intracellular domain was largely dispensable for induction of mitogenesis in the myeloid factor dependent cell line FDC-P1.[104,105] However, removal of sequences, which included the Box 2 region, markedly reduced but did not abrogate completely the capacity of GH to stimulate growth. For the G-CSF-R, a similar deletion series implicated both Box 1 and Box 2 in stimulation of mitogenesis,[107] while a follow-up study suggested that the Box 1 and Box 2 domains were sufficient to provide a mitogenic stimulus, but were insufficient to coordinate a differentiative signal.[106] The implication from this was that there were at least two independent signals emerging from the G-CSF-R, and that these were elements in the carboxy-terminal portion of this receptor that were necessary for the differentiation response, but not the mitogenic response.

Tyrosine Phosphorylation Sites

One of the first events that flows from the interaction of a cytokine receptor and its ligand is the tyrosine phosphorylation of one or more of the receptor chains[108-110] inter alia.

The biochemical analysis of phosphoproteins with the aim of defining phosphorylation sites is no longer in vogue. Thus, the information available on the tyrosine substrates borne by the intracellular domains of cytokine receptors is sparse to say the least. Rather, given that the likely role of such a phosphotyrosine residue might be the binding site of an SH2 domain, a common short cut has often been taken, and in vitro analyses of the SH2 binding potential of particular phosphotyrosines has been performed.[13,111] In these studies, candidate phosphotyrosine sites are represented by a series of short peptides derived from the intracellular domain of interest, and affinity chromotography with cellular extracts carried out using each of these peptides attached to beads. After careful washing of the beads, analysis of the proteins remaining bound is undertaken using Western blot analysis with antibodies against proteins likely to bind to these phosphotyrosine-containing motifs (such as SHC, PTP-SH1 or a member of the STAT family of transcription factors inter alia). The ability of a particular phosphopeptide to absorb out a given SH2-containing signal transduction molecule is widely regarded as being indicative of a functional role for this motif. Obviously, data generated in this way cannot be regarded as definitive; however, in the absence of detailed mapping data, they are the only data currently available.

Analysis of the gp130 and LIF intracellular domains has implicated a number of possible phosphorylation sites in the binding of second wave signal transduction molecules such as STAT3 (ARPF) and PTP-1D.[112] For these studies a series of deletion mutants, with increasingly smaller portions of the intracellular domain, were generated, and combined with the extracellular domain of the EGF receptor or Trk-C, so that ligand-induced homodimerization was sufficient to trigger autophosphorylation. A number of important tyrosine residues were located within the intracellular domain of the LIF receptor; one located at position 981 in the human LIF receptor appeared to be required for STAT3 binding, whereas one or both of two other candidates were required for the binding of PTP-1D. In the case of the gp130 intracellular domain, a tyrosine located at position 765 was required for the binding of STAT3, while the tyrosine at 759 appeared to bind PTP-1D.

Similar studies have been undertaken on the receptors for PRL,[113] EPO,[114] IFNα[115] in each case, it was noted that the deletion of particular tyrosines from the intracellular domains of these cytokines resulted in the loss of its capacity to bind SH2-containing molecules such as the STATs (see chapters 5 and 7 for more detail) and the p85 subunit of PI3'-kinase.[114]

Taken together, these data imply that there are a class of tyrosines located in the intracellular domains of cytokine receptors which have a function analogous to those found in the intracellular domains of RTKs, in that they bind SH2 domain-containing signal transduction molecules. The significance of this notion will become apparent in subsequent chapters.

CLASS I CYTOKINE RECEPTORS THAT HOMODIMERIZE

One feature upon which the cytokine receptor family can be categorized is their subunit structure. While all of the class II receptors and most of the class I receptors are composed of two or more subunits, a subset of the class I receptors (for example, the GH-R, the G-CSF-R and the EPO-R) appear to act as homodimers. For these receptors the binding of ligand to the extracellular domain appears to be sufficient to induce the dimerization of two identical polypeptide chains. Thus, cocrystallization of the GH-R with its ligand has shown that a single molecule of GH can bring about the dimerization of two GH-R chains.[102] The intracellular domain of each of these receptor chains is sufficient to bring about the coordination of the intracellular signal transduction molecules required to achieve the cellular response required.

GH Receptor

The receptor for growth hormone (GH) is perhaps the most well characterized receptor of the cytokine receptor family, its structure having been determined by X-ray crystallography.[102] The GH-R is made up of an extracellular domain of 246 amino acids linked via a 24 residue transmembrane to 350 amino acid intracellular domain. The GH-R is unique among the cytokine receptor family in that it does not possess a WSXWS motif in its hemopoietin domain, although a similar motif (YGEFS) is present in the appropriate location. The intracellular domain of the GH-R contains both a proline rich Box 1 and a Box 2 region.

PRL Receptor

The PRL-R is a close relative of the GH-R, being ~30% identical over their length. The rat PRL-R is found in a "long" form and a "short" form. Both forms bear the same extracellular domain of 234 amino acids long, however, the "short" form possesses an intracellular domain of 57 amino acids, whereas the "long" form possesses an intracellular domain of 357 amino acids. Only the "long" form of the PRL-R has a Box 1 and a Box 2 region, and it seems likely that this is the only form capable of initiating intracellular signal transduction events.

EPO Receptor

The mouse EPO-R is a protein of 507 amino acids, comprising an extracellular domain of 250 amino acids, with a transmembrane domain of 22 amino acids, and an intracellular domain of 235 amino acids. The extracellular domain is comprised of a single hemopoietin domain plus a single Ig domain. The intracellular domain bears both a Box 1 and Box 2 domain and is fully equipped for docking SH2 containing signal transduction proteins, with a number of tyrosine residues which become phosphorylated upon ligand stimulation (reviewed in ref. 116).

G-CSF Receptor

The G-CSF-R is present in small numbers (~300-1,000 sites per cell) in hemopoietic precursors of the neutrophil lineage and in mature neutrophilic granulocytes. Expression cloning[117] of this receptor revealed it to be one of the first described members of the cytokine receptor family. The extracellular domain of the G-CSF-R is composed of an N-terminal Ig-like domain, a single hemopoietin domain and three fibronectin type III repeats. The intracellular domain is large and possesses a Box 1, Box 2 and Box 3 regions. The closest relative of the G-CSF-R is gp130 of the IL-6/LIF/CNTF/OSM system (see below).

MULTIMERIC CYTOKINE RECEPTORS

As described above, a typical cytokine receptor chain is made up of a composite of a number of common domains, thus two hemopoietin domains may be juxtaposed to three FNIII repeats to generate a structure which will bind the appropriate ligand. As a general principle, those receptors composed of several polypeptide chains possess both a ligand binding chain (often referred to as the α-chain) and a signal transduction chain (often referred to as the β-chain). The multimeric nature of many of these receptors means that molecules with distinct hemopoietin domains may combine together to generate a single binding site for a particular cytokine, thus when an α-chain of, for example, the GM-CSF receptor, is brought together with the β_c chain of the GM-CSF/IL-3/IL-5 receptor system, a common binding pouch is formed so that the affinity modulation and signal transduction of the GM-CSF receptor signal transduction pathway is facilitated. The subunit composition of members of the IL-2 receptor family is both complex and may well vary from cell line to cell line, thus while the IL-2 receptor is composed of an IL-2 α-chain, an IL-2R β-chain and an IL-2R γ-chain, the β and γ components of this are shared with, for example, the IL-15 receptor.

The use of a common signal transduction chain by several of different ligand binding chains may serve one of a number of functions. It may serve as an affinity modulator generating high affinity binding sites from low affinity binding chains; it may integrate the biochemical consequences of a number of different ligands through the recruitment of common signal transduction pathways; or it may modulate the availability of receptor chains for other cytokines of the same class. Current notions make it seem likely that all three roles are undertaken by common receptor chains such as gp130 and β_c.

MULTIMERIC RECEPTORS THAT USE GP130: THE IL-6 FAMILY LIGAND-SPECIFIC BINDING COMPONENTS

IL-6-Rα

The IL-6 receptor α-chain (IL-6-Rα) is a transmembrane protein of 468 amino acids, possessing one HR domain and an Ig-like domain.[118] A secreted form of the IL-6 receptor has also been reported. The intracellular domain of IL-6-Rα is short (82 amino acids), does not possess either a Box 1 or Box 2 region, and is considered to be functional in a signal transduction sense only in combination with the gp130 signal transduction chain.[119]

LIF-Rα

The LIF-Rα is a low affinity binding chain for LIF.[120] The LIF-Rα is a 1097 amino acid transmembrane protein, with an extracellular domain of 789 amino acids and an intracellular domain of 238 amino acids. The LIF-Rα is a low affinity binding chain for LIF, requiring dimerization with gp130 before a high affinity LIF receptor is formed.[119-121] The LIF-Rα possesses an unusual constellation of domains in its extracellular domain, bearing as it does, a duplicated hemopoietin domain, separated by an Ig-like domain, as well as three FNIII repeats.[121] The intracellular domain possesses a Box 1 and a Box 2 homology domain,[121] and presumably recruits some of the signal transduction molecules required for the initiation of downstream events itself.

CNTF-Rα

The CNTF-Rα chain is a member of the hemopoietin domain family, possessing one HR domain and an Ig-like domain.[1] The CNTF-Rα is not a transmembrane protein itself, rather it is membrane associated via a glycosyl-phosphoinositol linkage.[1] The similar range of cellular responses elicited by both LIF and CNTF is thus due to the common use of signal transduction chains by both of these cytokines.[122,123] Conversely, the limited spectrum of biological activities elicited by CNTF is most likely to be due to the restricted pattern of expression of the CNTF-R. This fact has made feasible the intriguing possibility of making "designer cytokines" wherein CNTF-Rα can be targeted to particular tissue whereupon a phenotopic "LIF" response can be elicited by CNTF administration (G. Yarcopoulos, personal communication).

IL-11-Rα

The murine IL-11 receptor was cloned by dint of the use of degenerate oligonucleotides derived from the WSXWS motif common to almost all cytokine receptors.[124] The IL-11-Rα is composed of a single hemopoietin domain and a single Ig-like domain. The extracellular domain is 367 amino acids in length, while the intracellular domain is only 39 amino acids long. The IL-11-Rα requires gp130 for effective signaling, and for the formation of a high affinity binding site. Thus the IL-11 receptor is a typical member of the class of cytokine receptors which require common β-chain gp130 for function.

GP130: The Common Signal Transduction Chain

The common β-chain of the receptors for the IL-6/IL-11/LIF/CNTF/OSM family (gp130) is an 130 kDa transmembrane glycoprotein, originally described as a component of the IL-6 receptor. Human gp130 possesses a 597 amino acid extracellular binding domain, comprising, from N-terminus to C-terminus, an Ig-domain, a single hemopoietin domain, and three FNIII domains.[121] The intracellular domain of gp130 is 277 amino acids in length, and contains both a Box 1 and a Box 2 region.

Cytokine Receptors Made Up of the GP130 and a Specific Ligand Binding Chain

The unexpected discovery that the CNTF-Rα was related to IL-6-Rα,[1] led to the notion that this ligand also required a β-chain of some kind to generate a functional receptor. In short order, the search for a β-chain for CNTF, as well as LIF and OSM, led to the discovery that gp130 became phosphorylated upon tyrosine residues in response to stimulation with each of these ligands. Moreover, co-immunoprecipitation of CNTF and LIF cross-linked to gp130 suggested that this β-chain was indeed common to both receptors.

The picture became more complicated still with the emergence of a second "β-chain" for the CNTF and OSM.[125] In fact, the second CNTF-Rβ chain emerged as none other than the so-called LIF-Rα chain[120] resulting in a shift in nomenclature and the rehash of LIF-Rα as LIF-Rβ. Thus, the CNTF receptor is composed of three chains: CNTF-Rα the 70 kDa binding chain, and the two signal transducing β-chains, LIF-Rβ and gp130. On the other hand, LIF did not appear to require an α-chain (indeed, LIF-Rβ was the α-chain), but nonetheless required gp130 for function.[126] Clearly, the biochemistry of the IL-6/IL-11/LIF/CNTF/OSM receptor system is immensely complicated. Figure 6 shows the extent of our ideas of the composition of each of the receptors for this family of cytokines.

Multimeric Receptors That Use β_C: The IL-3/GM-CSF/IL-5 Family

The IL-3/GM-CSF/IL-5 family of cytokines possess overlapping activities upon cells of hemopoietic origin. In particular, these cytokines play a role in the differentiation, proliferation and activation of mature cell function of a number of myeloid lineages. Thus, using mouse bone-marrow as a source of hemopoietic cell precursors, murine IL-3[82] (known also as multi-CSF) is capable of supporting the in vitro development of mixed myeloid/erythroid, macrophage, neutrophil, eosinophil and basophil colonies, whereas GM-CSF supports the development of granulocyte, macrophage and mixed granulocyte/macrophage colonies, and IL-5 develops most eosinophil colonies. The hierarchical nature of the hemopoietic differentiation program, and the overlapping activities and built-in degeneracy of the factors themselves, has been in itself an intriguing biological puzzle for nearly two decades. Since the successful cloning and characterization of the receptors for each of these cytokines, an analysis of the mechanics of the signaling events which stem from their ligand mediated activation has become increasingly possible. The alpha chains (ligand binding chains) of the receptors for IL-3 (IL-3-Rα), GM-CSF (GM-CSF-Rα)[127] and IL-5 (IL-5-Rα)[128,129] have each been molecularly cloned. A comparison of the amino acid sequences of these three receptors underlines the familial relationship deduced for the three ligands themselves. Further, a common β component, apparently required for a range of signal transduction events has been discovered, and a plausible, albeit complex, model to explain the related functions of the three ligands has been developed. Each of the components of this ligand receptor family will be discussed in detail.

LIGAND-SPECIFIC BINDING COMPONENTS

IL-3 Rα, GM-CSF-Rα and IL-5-Rα

The α-chain of the human IL-3, human GM-CSF[127] and human IL-5[128,129] receptors are 396, 400 and 420 amino acids in length, respectively. The extracellular domains of these receptors are similar in size (315 amino acids for IL-3-R; 298 amino acids for GM-CSF-R; and 322 amino acids for IL-5-R), and each bears a short intracellular domain of around 50 amino acids (41 amino acids for IL-3-R; 54 amino acids for GM-CSF-R; and 58 amino acids for IL-5-R). In common with the receptors for GM-CSF and IL-5, the IL-3 receptor has a characteristic 200 amino acid cytokine receptor extracellular domain module containing four

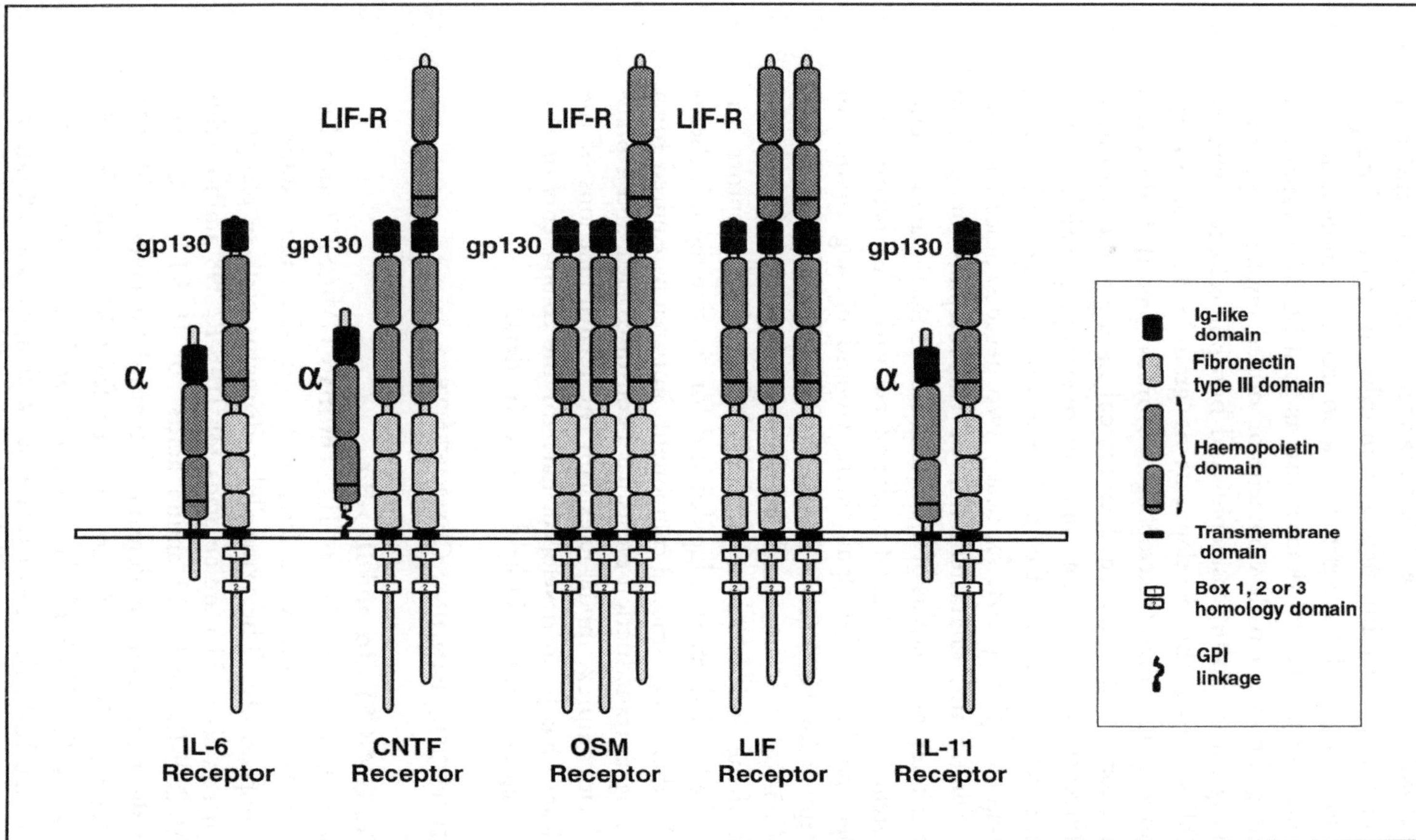

Fig. 6. Schematic of the arrangement of domain structure of heterodimerizing cytokine receptors that use the gp130 chain. The domain structures of the IL-6/CNTF/OSM/LIF and IL-11 receptors are shown. Each these receptors, with the exception of OSM, are composed of an α-chain, combined with the same β-chain, the gp130. The OSM receptor is composed of two gp130 proteins combined with a LIF receptor chain. The CNTF receptor is tethered to the membrane via a GPI linkage.

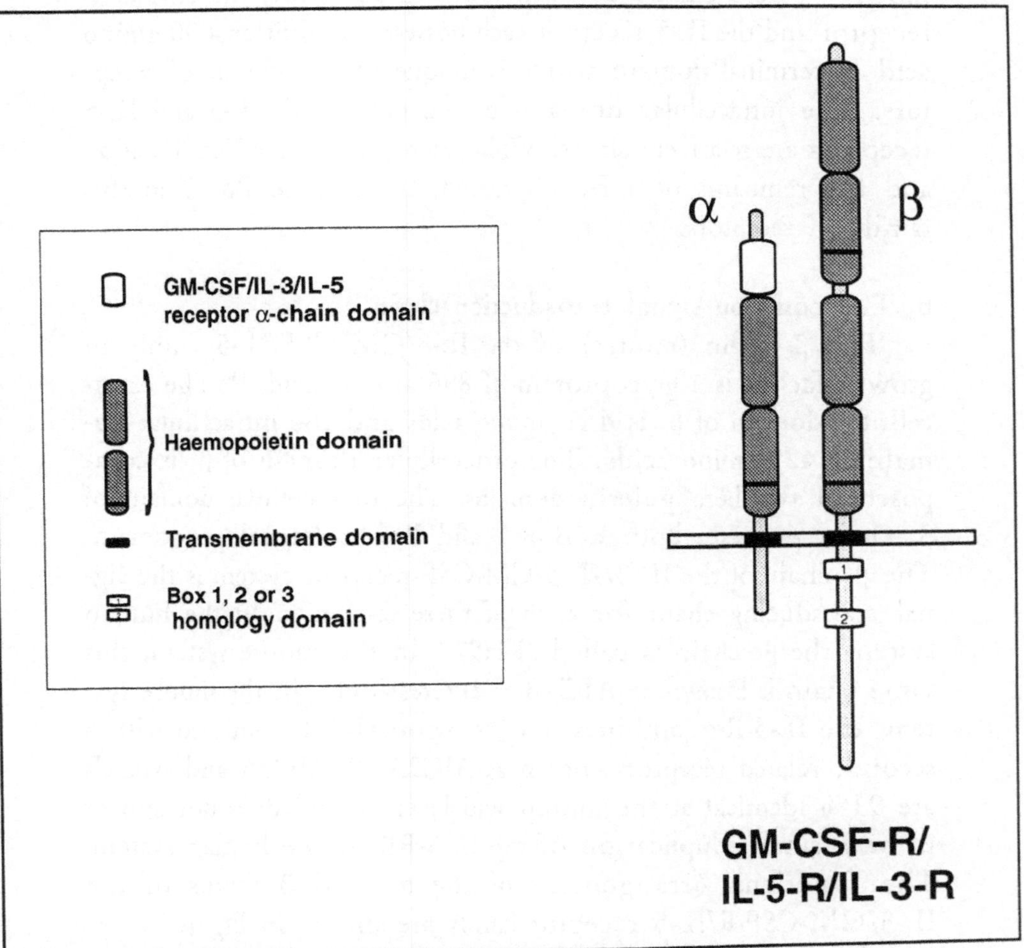

Fig. 7. Schematic of the arrangement of domain structure of heterodimerizing cytokine receptors that use the β_c chain. The domain structures of the GM-CSF/IL-3/IL-5 receptors are shown. Each these receptors are composed of an α-chain possessing a single "hemopoietin domain", combined with the same β-chain, the AIC2B, which bears two hemopoietin domains.

conserved cysteine residues and a membrane promixal WSXWS motif.[79] The α-chains of both the IL-3 receptor, the GM-CSF receptor and the IL-5 receptor each possess an additional 90 amino acid N-terminal domain which is unique to this family of receptors. The intracellular domains of the GM-CSF, IL-3 and IL-5 receptors are relatively short. While each possesses a Box 1 region and the remnants of a Box 2 region, there is no Box 3 in this family of receptors.

β_c: The common signal transduction chain

The β_c chain (mouse) of the IL-3/GM-CSF/IL-5 family of growth factors is a glycoprotein of 896 amino acids.[130] The extracellular domain of β_c is 441 amino acids and the intracellular domain is 429 amino acids. The extracellular domain of β_c is composed of two hemopoietin domains. The intracellular domain of β_c chain possesses both a Box 1 and Box 2 homology domain. The β_c chain of the IL-3/IL-5/GM-CSF-receptor system is the signal transducing chain for each of these α-chains. In the human system the βc chain is called "KH97." In the mouse system, this same chain is known as AIC2B.[130] Interestingly, in the mouse system, the IL-3-Rα combines not just with AIC2B, but also with a second, related receptor known as AIC2A.[131] AIC2A and AIC2B are 91% identical at the amino acid level. There does not appear to be a similar duplication of the IL-3-Rβ in the human system. The functional arrangement of the α- and β-chains of the IL-3/GM-CSF-R/IL-5 receptor family are shown in Figure 7.

RECEPTORS WHICH USE THE COMMON CHAIN γ_C

The receptors for IL-2,[132-134] IL-4,[135] IL-7,[136] IL-9,[137] IL-13[138] and IL-15,[139] are expressed upon lymphocytes of both B cell and T cell lineages. In addition, receptors for IL-4 and IL-13 are found upon a range of other cell types. Each of these cytokine receptors share a common signaling chain known as γ_c,[134,140-143] although the involvement of this protein in the signaling of IL-9 and IL-13 remains contentious. Thus, each of these ligands binds to its own multimeric receptor complex composed of an α-chain, β-chain in

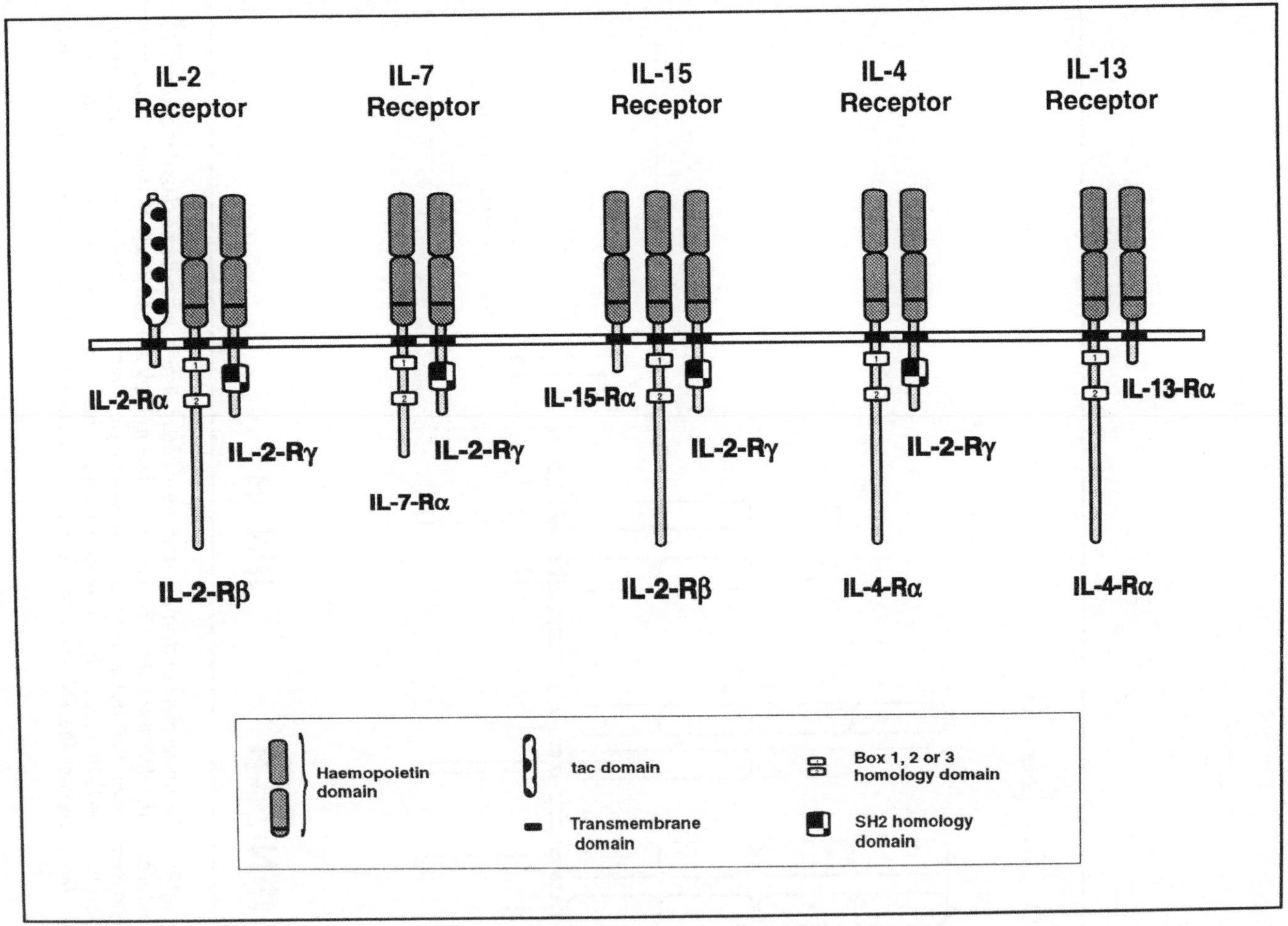

Fig. 8. Schematic of the arrangement of domain structure of heterodimerizing cytokine receptors that use the IL-2 Rγ chain. The domain structures of the IL-2/IL-7/IL-15/IL-4 and IL-13 receptors are shown. Each these receptors, with the exception of the IL-13, are composed of an α-chain, combined with the same "β-chain", the IL-2 Rγ. Th IL-13 receptor is composed of the IL-4-Rα and a specific IL-13 α chain.

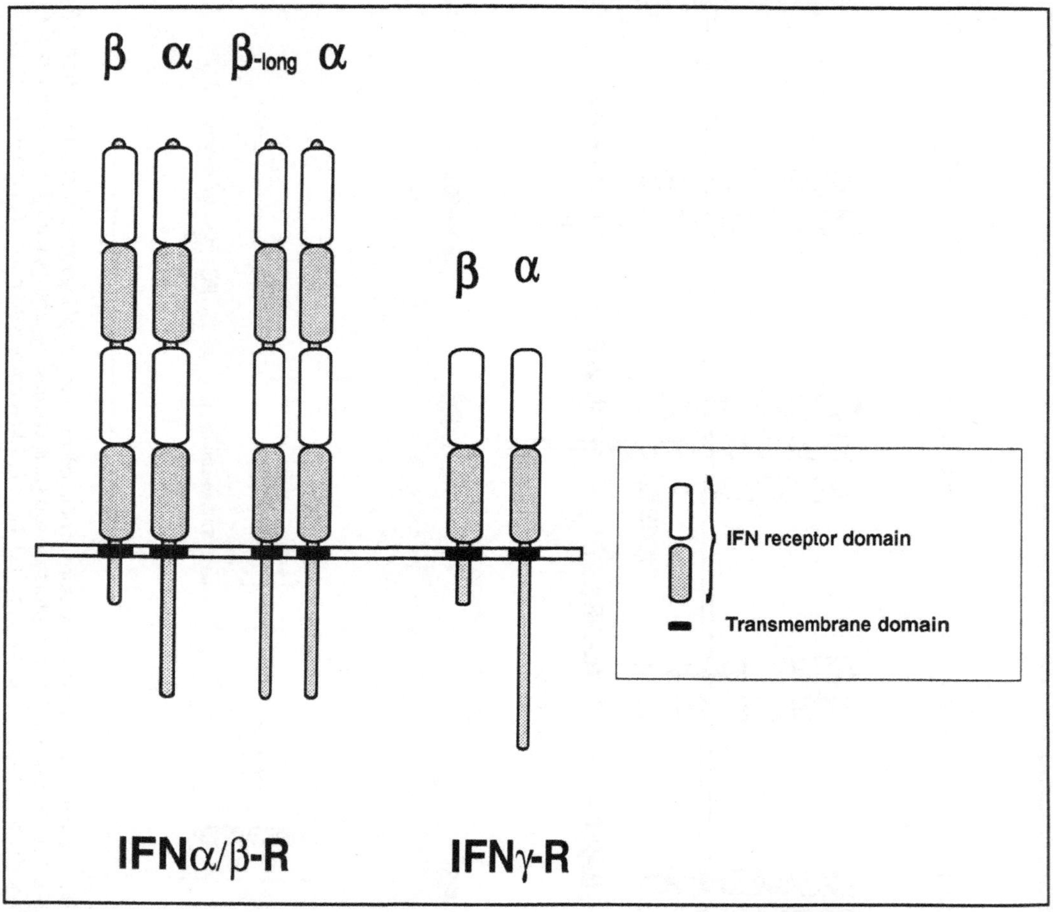

Fig. 9. Schematic of the arrangement of domain structure of heterodimerizing cytokine class II receptors. The domain structures of the IFNα/β and IFNγ receptors are shown. They are composed of at least two chains. The IFNα/β possesses two "IFN" domains in its extracellular region, whereas, the IFNγ receptor possesses a single IFN domain. The IL-10 receptor is of unknown receptor chain organization, however, it is structurally related to the IFN receptors. The IFN receptors do not possess a Box 1 or Box 2 domain in their intracellular regions.

addition to γ_c. The composition of each of the receptors for these cytokines is shown in Figure 8.

α-CHAINS AND β-CHAINS

The α-chains (ligand binding chains) of IL-2,[132] IL-7,[136] IL-9[137] and IL-15[139] have all been characterized by dint of cloning and sequencing. At present, the ligand binding chain for IL-13 has not been molecularly cloned, although some preliminary characterization of the subunit composition of its receptor has been carried out.[138,144]

IL-2-Rα

The human IL-2 Rα (also known as p55 or Tac) is a 272 amino acid transmembrane protein.[132] The extracellular domain is 219 amino acids in length, and is comprised of a single "Tac" domain, a feature which is not shared by any other cytokine receptor. The intracellular domain is very short—only 13 amino acids—and does not appear to possess any signaling capability.

IL-2-Rβ

The human IL-2-Rβ chain is a 551 amino acid polypeptide, comprising an extracellular domain of 214 amino acids, and an intracellular domain 286 amino acids.[133] The extracellular domain is similar to the IL-7 receptor, and is composed of a single hemopoietin domain. The intracellular portion of the IL-2-Rβ bears a Box 1 domain and an acidic Box 2 domain.

IL-4-Rα

The human IL-4-Rα is an 810 amino acid transmembrane protein composed of an extracellular domain of 208 amino acids, and an intracellular domain of 553 amino acids.[135] The extracellular domain is made up of a single hemopoietin domain. The intracellular domain bears both a Box 1 and a Box 2 region.

IL-7-Rα

The human IL-7-Rα is a 459 amino acid transmembrane protein composed of an extracellular domain of 219 amino acids, and an intracellular domain of 195 amino acids.[136] The extracellular domain is made up of a single hemopoietin domain.

IL-9-Rα

The mouse IL-9-Rα is a transmembrane protein of 468 amino acids, composed of an extracellular domain of 233 amino acids, and an intracellular domain of approximately 180 amino acids.[137] The extracellular domain possesses a single hemopoietin domain. The intracellular domain bears both a Box 1 region and a Box 2 region.

The γ_C Chain

For the IL-2 receptor, the α-chain is a high affinity binding chain, while both the β-chain and the γ_c chain coordinate differing components of the signaling pathways downstream of the receptor.[140,145-147] The γ_c is a 369 amino acid transmembrane protein, with an extracellular domain of 232 amino acids, and an intracellular domain of 86 amino acids.[134] The extracellular domain is comprised of a single hemopoietin domain. The intracellular domain contains an "SH2 homology" domain. The importance of the γ_c chain is underlined by its implication in the etiology of the clinical syndrome known as X-linked severe combined immunodeficiency syndrome (X-SCID).[148]

In certain patients with this disease, there are deletions in the intracellular domain of the γ_c chain which render signaling impossible.[148] Interestingly, mice which are homozygous-null for IL-2 do not display the same clinical features of X-SCID patients, suggesting that the lack of γ_c function has a broader impact than simple loss of signaling in the IL-2 pathway.

It seems likely that the primary role of γ_c in each of the multimeric receptors described, is the recruitment of JAK3 to the receptor complex.[149] There is evidence of a direct interaction of JAK3 with γ_c,[149] moreover, X-SCID patients are unable to activate JAK3 in response to IL-2 interaction.

The compositions of the IL-2, IL-4, IL-7, IL-13 and IL-15 receptors are shown in Figure 8.

THE CLASS II CYTOKINE RECEPTORS

IFNα/β-RECEPTOR

The class II cytokine receptor family is composed of the IFNα-R,[150,151] IFNγ-R[152-154] and IL-10-R.[155] IFNα and IFNβ are grouped together as the type I interferons, whereas, the mitogen induced IFNγ is a type II interferon. For the IFNα-R, there are at least two (probably three) receptor components required for signaling, namely IFNA-Rα[151] and IFNA-Rβ.[150] The third component is not yet characterized with respect to sequence, but has been inferred from cross-linking studies. The IFNA-Rα is a glycoprotein of 557 amino acids, possessing an extracellular domain of 409 amino acids (composed of two SD100 domains), and an intracellular domain of 100 amino acids. The IFNA-Rα chain, with a molecular mass of 102 kDa, clearly has a role in IFNα/β binding, however, it seems likely that the intracellular domain has a role to play in intracellular signaling by recruiting JAK1 to the IFNA-R complex. The IFNA-Rβ is perhaps less well understood. This chain's role in ligand binding is not yet fully understood (it binds IFNα, but not IFNβ, for example), but it clearly plays an important role in signal transduction, since it is rapidly phosphorylated in response to IFNα and IFNβ.

The IFNA-Rα chain is found in two forms known as the "short" and "long" form.[156] The short form is truncated by differential splicing, and is missing the carboxy-terminal 200 amino acids with respect to the long form. Interestingly, the short form is unable to complement one of the IFNα/β somatic cell signaling mutants[156] (U5) (see chapter 6), while the long form complements the defect. This suggests that there are elements on the C-terminal portion of the IFNA-Rα which are required for signaling.

IFNγ-RECEPTOR

The IFNγ-R is composed of two chains, IFNγ-Rα[152] and IFNγ-Rβ.[153,154] Both components are characterized in terms of their primary sequence. The IFNγ-Rα is a glycoprotein of 472 amino acids, possessing an extracellular domain of 228 amino acids (composed of a single SD100 domains), and an intracellular domain of 221 amino acids. The α-chain possesses a JAK binding domain

KLEP and a tyrosine phosphorylation site located at position 440 in the human sequence. The IFNγ-Rβ is a glycoprotein of 337 amino acids, with an extracellular domain of 226 amino acids (composed again of a single SD100 domains), and an intracellular domain of 66 amino acids. There are no known JAK binding sites or tyrosine phosphorylation sites assigned to date.

IL-10 Receptor

IL-10 is a cytokine produced by activated T cells, B cells, keratinocytes and monocytes. Both mouse and human IL-10 have pleiotropic effects on a range of cell types, including stimulation of proliferation and differentiation of B cells, T cells and mast cells, and the inhibition of cytokine synthesis in activated T cells. cDNAs for both the murine[155] and human[157] IL-10 receptors have been cloned and characterized, and it is now apparent that this receptor is a member of the class II cytokine receptor family. The IL-10 receptor is an approximately 110 kDa glycoprotein which is capable of binding IL-10 with high affinity. At present, a second chain for the IL-10 receptor has not been described. However, many of the effects of IFNγ are inhibited by IL-10, suggesting that there is some interaction between the two receptor systems. It may be that these two receptors share one or more components, in a manner akin to that which prevails for the IL-2 receptor family or the IL-3 receptor family. Alternatively, there may be some cross-talk between the signal transduction pathways which lie downstream.

SIGNALING EVENTS DOWNSTREAM OF CYTOKINE RECEPTORS

As with the RTKs, one of the first intracellular biochemical events which follows the interaction of a cytokine receptor with its cognate ligand is the tyrosine phosphorylation of one of the receptor chains[108-110] inter alia. In addition to the receptor chain, a cytokine-specific pattern of tyrosine phosphorylated proteins is produced[158], suggesting that the events downstream of cytokine receptors is thematically the same as those which lie downstream of RTKs. Clearly, one of the major targets of cytokine signaling is ras. Activation of ras has been observed in response to IL-2,[159,160]

IL-3,[159,160] IL-5,[161] GM-CSF[161] and EPO.[162] Finally, MAPK activation appears to be a consequence of many cytokines, for example GM-CSF, IL-3 and IL-5[163] all appear to stimulate MAPK activity. Thus, the themes of tyrosine phosphorylation and activation of the ras/MAPK pathway are held in common by both RTKs and cytokine receptors. Whereas for the RTKs, the intrinsic PTK domain is the obvious candidate for providing the lynch-pin kinase activity which both initiates and coordinates the cascade of signal transduction leading to MAPK activity, for the cytokine receptors the issue is less clear. Several nonreceptor PTKs are candidates for recruitment to the receptor. Thus, *FES*,[164] and members of the *SRC* family of PTKs have been mooted as candidates for the provision of PTK activity to the non-PTK cytokine receptors. More recently, however, members of the JAK family of PTKs[165-169] have proven to be the most likely candidates; the remainder of this book will focus upon the evidence, both genetic and biochemical, that members of this family are pivotal to the signal transduction processes which lie downstream.

CHAPTER 3

The JAK Family of Protein Tyrosine Kinases

The JAK family of nonreceptor PTKs were discovered by application of PCR, using degenerate oligonucleotide primers based on highly conserved elements located within the catalytic domain of all PTKs.[165-167,170,171] At the time of discovery they were but two sequences from a multitude of previously unknown PTK-related sequences emerging from this PCR-based screen (earning the laboratory code-name of *Just Another Kinase; or JAK)*. Ultimately, the discovery that these proteins possess two kinase-related domains suggested a more learned justification of the acronym 'JAK', as the *JAnus Kinases*, named after the Roman god Janus, who possessed two faces.[172] The contemporaneous discovery of Tyk2,[168,169] a third member of this class of PTKs, and the more recent discovery of JAK3,[173-176] complete a cast of four JAK family members in the human (and mouse) genome. More recently, a *Drosophila melanogaster* family member (named, with typical fly community imagination, *hopscotch)* has been discovered,[177] and a piscine homologue of JAK1, (known as F-JAK and only as an unpublished sequence from its accession number in the Genbank database), has been isolated from carp *(Cyprius carpio)*. A further two piscine JAK family members have been isolated from the Zebrafish *(Danio rerio)*, and appear to be JAK1 and JAK2 homologues (Oates, Irvine and Wilks, unpublished data).

Thus, the JAKs appear to be an evolutionarily well conserved branch of the PTK family; not as extensive as the *SRC* family (with 14 known members[7]); nor as small as the *abl* or *fes* families (with two known mammalian members each; *abl* and *arg*; and *fes* and *fer*, respectively[7]). In order to place these PTKs in the context of their role in cytokine signal transduction, we will deal in detail with the basic facts about mammalian members of the family, namely JAK1, JAK2, Tyk2 and JAK3. A detailed analysis of *hopscotch* and its oncogenic form *Tum^l* appears in chapter 8.

cDNA ISOLATION AND CHARACTERIZATION

JAK1

JAK1 cDNAs were first isolated from a murine myelomonocytic cDNA preparation by dint of application of PCR using oligonucleotide primers based upon PTK homology domains.[165] Full-length cDNAs were subsequently obtained from a human foreskin fibroblast library.[167] The longest open reading frame encompassed two potential in-frame ATG start codons. The first putative start codon was embedded in a particularly poor Kozak consensus[178] (-TAAATGCAG-), and the second conformed well to the optimal consensus sequence GCCATGGCT. However, comparison with the mouse JAK1 sequence (Harpur and Wilks, unpublished[179]), suggested that the first ATG was correct, since the mouse and human amino acid sequences predicted from these two sequences are identical. The open reading frame of 3465 bp predicted a human JAK1 protein of 1155 amino acids with a calculated molecular mass of 132 kDa. The mouse protein is two amino acids shorter at 1153 amino acids.

JAK2

JAK2 was, like its close relative JAK1, discovered by application of PCR using oligonucleotide primers based upon PTK homology domains.[165,167] cDNAs covering the entire length of mouse JAK2 were ultimately isolated from a number of libraries.[166] An open reading frame of 3411 bp predicted that mouse JAK2 encoded 1137 amino acids with a calculated molecular mass of 130 kDa. At present, no full-length human JAK2 cDNAs have been described.

TYK2

Tyk2 was discovered by low stringency hybridization screening of a human lymphoid cDNA library with the PTK domain of the *c-fms* receptor.[168] Subsequent screening of a cDNA library from the human T cell lymphoma cell line, PEER, identified a full length cDNA with an open reading frame of 3561 bp. Interestingly, six additional ATG start codons followed by in-frame stop codons were present within the 5'-untranslated region. The longest open reading frame encoded a polypeptide of 1187 amino acids with a predicted molecular mass of 134 kDa.[180]

JAK3

The fourth member of the JAK family to be uncovered (JAK3) was published almost simultaneously by four independent groups.[173-176] Human JAK3 (previously termed L-JAK due to its initial identification in leucocytes) cDNAs were cloned from natural killer (NK) cells,[173] and the encoded protein had a predicted molecular mass of 125 kDa. Murine JAK3, isolated from a B cell library, had an open reading frame encoding a protein of 1099 amino acids (122 kDa). JAK3 is thus the smallest of the JAK family of PTKs, however, it retains all of the features characteristic of this class of PTKs.

Several interesting splice variants of JAK3 have been identified in human mammary cells and human monocytes.[181] The mammary variant, designated JAK3B (JAK3-Breast), is apparently generated by failure to splice across an intron located in the portion of the JAK3 gene coding for motif XI[4,6] of the PTK domain. The resulting read-through product is thus lacking a functional PTK domain. That a protein was indeed encoded by this transcript was demonstrated by immunoprecipitation of a short form of JAK3 in these cells. JAK3B appeared to have a more extensive expression pattern than the "full-length" JAK3. The functional significance of these truncated JAK3 proteins is unknown at present.

EXPRESSION

Human and mouse JAK1 mRNA are expressed as a 5.4 kb transcript and have been found in all tissues and cell lines examined.[167] Fine resolution studies with mouse JAK1 by in situ hybridization and immunohistochemistry demonstrated that the expression of JAK1 mRNA and protein was localized to the inner section of the inner nuclear layer (presumed to be in amacrine cells) and the retinal ganglion cell layer in the mouse eye.[179] Within the mouse brain, JAK1 is expressed in the olfactory bulb and selected cells in the hypothalamus (including the dorsal premammillary nucleus, the compact dorsomedial hypothalamic nucleus, the subincertal nucleus and the lateral magnocellular hypothalamic nucleus).[179] Thus, while mRNA populations taken from entire tissues such as the brain, suggest ubiquitous expression, it may be that particular cell types within a tissue may express either more or less JAK1.

JAK2 also appears to be expressed in all tissues and cell lines examined,[167] and is encoded by two transcripts of 5.3 kb and 5.0 kb.[167] Both transcripts are large enough to encode a full-length JAK2 protein, and the significance of the existence of two transcripts is not known at present.

Northern blot analysis of Tyk2 revealed the wide distribution of a 4.4 kb transcript in a number of hematopoietic and nonhematopoietic cell lines examined (including a selection of T and B cell lines and epithelial and epidermal cell lines).

While the expression patterns of JAK1, JAK2 and Tyk2 are apparently ubiquitous distribution of JAK3 expression is somewhat more restricted. At the level of sensitivity of Northern blot analysis there are no transcripts present in fibroblasts, while abundant expression of a 4.0 kb JAK3 cDNA has been detected in IL-2-dependent cytotoxic T cells (CTLL) and myeloid cells,[175] and also in rat hippocampal neurons.[182] The alternatively spliced form of JAK3 described in human mammary carcinoma cell lines, appears to be more widely expressed than the transcript encoding the full-length JAK3 protein found in cells of hematopoietic and epithelial origin.[181]

CHROMOSOMAL LOCALIZATION

The chromosomal localization of human JAK1 (1p31-3),[183] JAK2 (9p24),[183] Tyk2 (19p13.2)[169] and JAK3 (4q31) have each been determined by cytogenetic means. The murine homologues have been genetically mapped by the use of recombinant inbred

(RI) strains. The two mouse genes that have been mapped are largely located in regions of synteny with the human chromosomal loci. Thus JAK1 is located on mouse chromosome 4 and JAK2 is on chromosome 19.[184] These data are presented in Table 3.

Table 3. Chromosomal localization of members of the JAK and STAT families

	MW	*Chromosomal Location*	*Reference*
JAK1	132 KD	1p31-3 (Human) 4 (Mouse)	183 184
JAK2	128KD	9p23-p24(Human) 19 (Mouse)	183 184
JAK3	120KD	4q31	176
Tyk2	140KD	19p13.2	180
STAT1	91/84KD	2q32-q34(Human) Chr 1 (Mouse)	(Predicted) Copeland et al, in press
STAT2	113KD	12q13-q14.1(Human) Chr 10 (Mouse)	(Predicted) Copeland et al, in press
STAT3	93/90KD	17q11.2-q22 (Human) Chr 11 (Mouse)	(Predicted) Copeland et al, in press
STAT4	89KD	2q32-q34 (Human) Chr 1 (Mouse)	(Predicted) Copeland et al, in press
STAT5	92/84KD	17q11.2-q22 (Human) Chr 11 (Mouse)	(Predicted) Copeland et al, in press
STAT6	100KD	12q13-q14.1(Human) Chr 10 (Mouse)	(Predicted) Copeland et al, in press

The mouse and human locations for the genes encoding the JAKs and STATs are shown.

JAK HOMOLOGY (JH) DOMAINS

A direct comparison of the four mammalian JAK family members revealed the presence of seven highly conserved domains.[166] In seeking a nomenclature for the highly conserved domains characteristic of this family of PTKs, our classification has been guided by the approach of Pawson and coworkers[185] in their treatment of the *SRC* homology (SH) domains. The domains have been enumerated accordingly with most C-terminal homology domain designated JAK Homology domain 1 (JH1). The next domain N-terminal to JH1 is the kinase-related domain, designated here the JH2 domain. Each domain is then enumerated up to the JH7 located at the N-terminus (refer to Fig. 10 for a schematic representation of this nomenclature). The high degree of conservation of these JAK homology (JH) domains suggests that they are each likely to play an important role in the cellular processes in which these proteins operate. However, the boundaries of the JAK homology domains are arbitrary, and may or may not define functional domains, nonetheless, their delineation is a useful device to aid the consideration of the overall structural similarity of this class of proteins.

THE PTK DOMAIN

The feature most characteristic of the JAK family of PTKs is the possession of two kinase-related domains (JH1 and JH2)[167] (Figs. 10 and 11). The putative PTK domain of JAK1 (JH1) contains highly conserved motifs typical of PTK domains, including the presence of a tyrosine residue at position 1022 located 11 residues C-terminal to subdomain VII, that is considered diagnostic of membership of the tyrosine-specific class of protein kinases. Alignment of the human JAK1 PTK domain (255 amino acids), with other members of the PTK class of proteins revealed homology with other functional PTKs (for example, 28% identity with *c-fes*[164] and 37% homology to TRK[186]). The JH1 domains of each of the JAK family members possess an interesting idiosyncrasy within the highly conserved subdomain VIII motif (residues 1032 to 1039 in JAK1, Fig. 10) that is believed to lie close to the active site, and define substrate specificity.[164] The phenylalanine and tyrosine residues flanking the conserved tryptophan in this motif are unique to the JAK family of PTKs.[166,167,173,180]

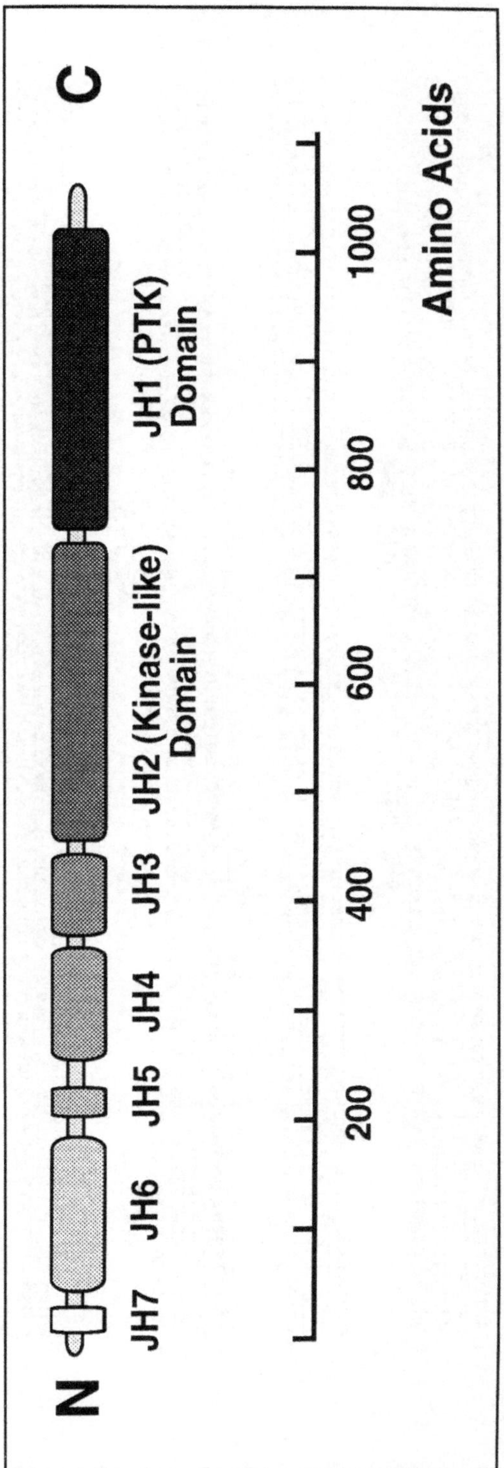

Fig. 10. Structure of JAK family members. The seven JAK homology domains (JH domains) are shown as shaded boxes. JH1 is the PTK domain, while JH2 is the kinase-like domain, that is characteristic of this class of PTKs. The drawing is approximately to scale.

Fig. 11. Comparison of the sequences of the four mammalian JAK family members JAK1, JAK2, Tyk2 and JAK3. The amino acid sequences (in one letter code) of the four mammalian JAK family members is shown. The location of the boundaries between the JH domains is illustrated by a bracket.

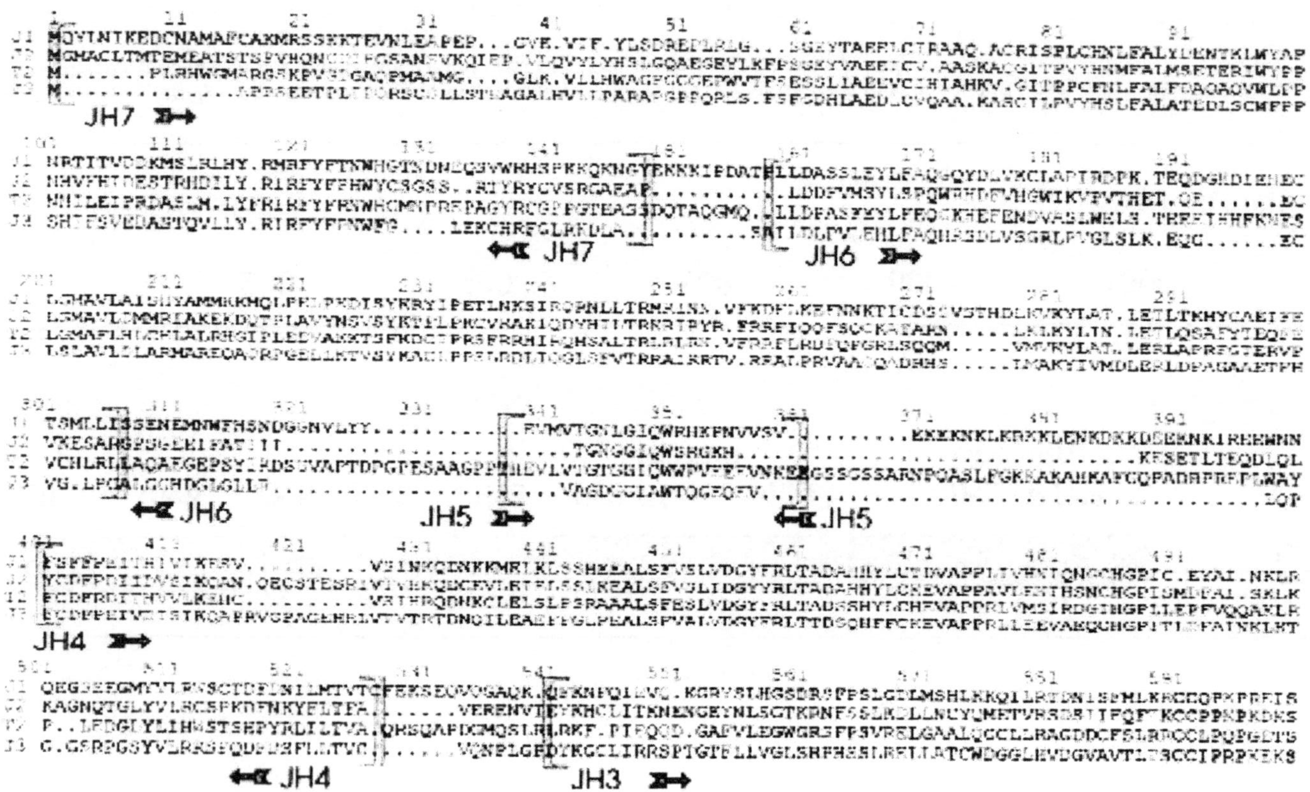

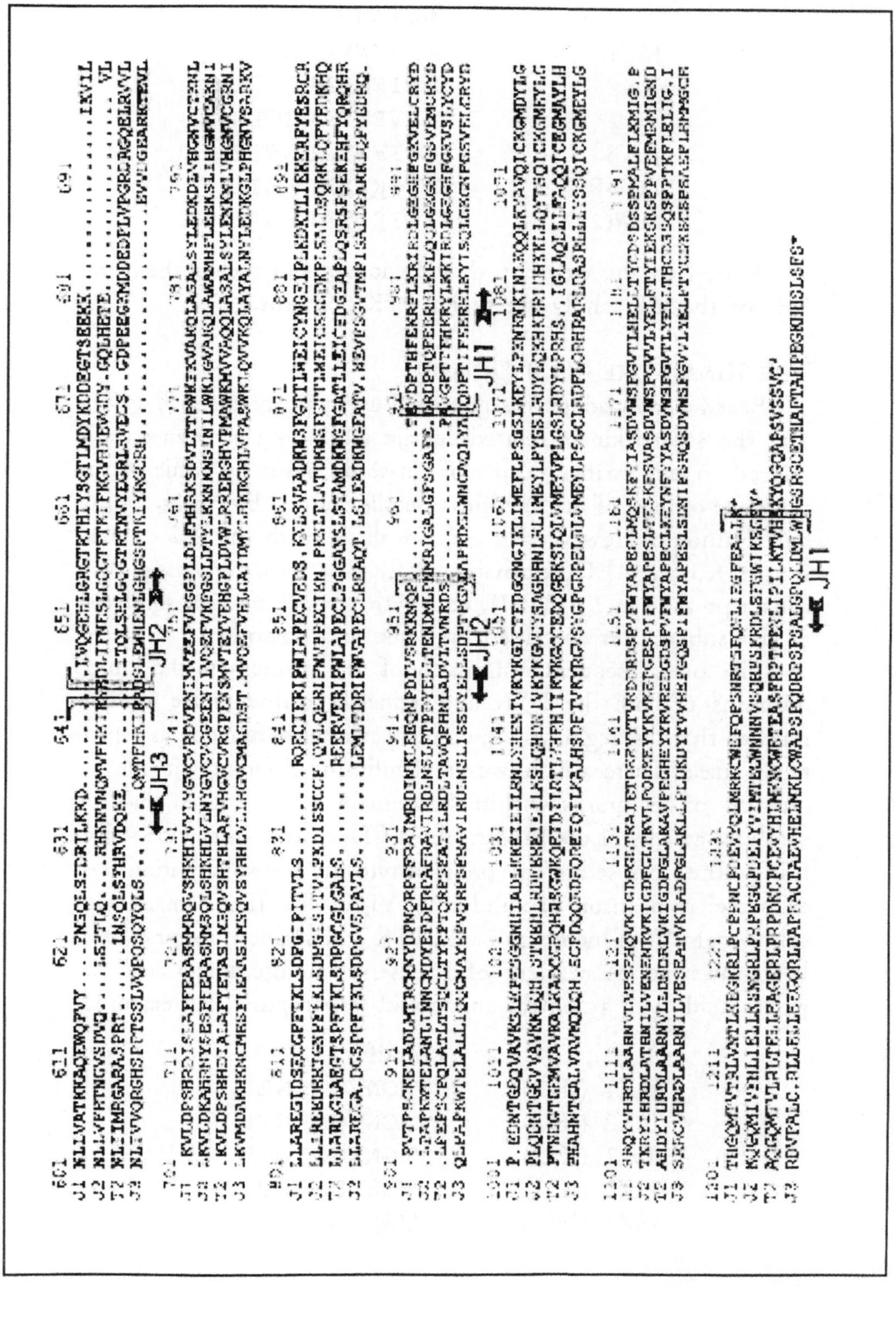
JH3
JH2
JH1

	Motif VIII
JAK1	DSPVFWYAPECLI
JAK2	ESPIFWYAPESLT
Tyk2	DSPVFWYAPECLK
JAK3	QSPIFWYAPESLS
EGF-R	KVPIKWMALESIL
c-*SRC*	KFPIKWTAPEAAL

Aside from this element, the JH1 domains of each of the members of the JAK family are typical PTK domains.

THE KINASE-LIKE (JH2) DOMAIN

Based upon cladograms generated using programs such as *Pile Up*, the second kinase-related domain (JH2) is clearly ancestrally related to the broader family of kinase domains by virtue of the presence of most of the key kinase motifs defined by Hanks, Quinn and Hunter.[4,6] However, in order to distinguish the JH2 domain motifs from the PTK domain motifs, they have been assigned a subscript a, (e.g., I_a, II_a, III_a, etc.) with respect to their similarity to the subdomains described by Hanks and coworkers.[4]

The overall sequence similarity of this domain to the kinase domains of both the PTK and serine/threonine kinase families implies that this region of the protein may also function as a protein kinase. There are, however, significant differences in the sequences of key motifs within this domain which suggest that the catalytic activity of the kinase-related (JH2) domain may be something other than serine/threonine or tyrosine phosphorylation. For example, comparison of subdomain VI_a of the JH2 domain with subdomain VI of members of the PTK and serine/threonine families shows the replacement of a conserved acidic amino acid (aspartic acid) with a neutral amino acid (asparagine). For example,

	Motif VI_a
JAK1 (JH2)	VHGNVCTKNLL
JAK2 (JH2)	IHGNVCAKNIL
Tyk2 (JH2)	VHGNVCGRNIL
JAK3 (JH2)	PHGNVSARKVL
JAK1 (JH1)	VHRDLAARNVL

EGF-R	VHRDLAARNVL
cAMPKα	IYRDLKPENLL

Further, while there is conservation of subdomain VII_a with respect to the equivalent motif in the other kinase families, the normally invariant D-F-G sequence of the PTK and serine/threonine families (motif VII) is replaced by the sequence D-P-G in motif VII_a of the JH2 domain. The conservation of the precise sequence of subdomain VI in the protein kinase subfamilies appears to correlate with the substrate specificity of the kinase,[4] and thus it is possible that this domain within the members of the JAK family of PTKs, may exhibit a substrate specificity other than that previously observed for other protein kinases.

A further sequence anomaly that may suggest a substrate variation lies within the putative ATP-binding site in the kinase-related domain (subdomain I_a). This domain consists of the absolutely conserved -GXGXXG- in all PTKs described to date. However, in all known JAK family members, subdomain I_a is replaced with -GXGXXT-. This glycine motif has now been defined as the ATP-binding site, with the first two glycine residues thought to bend around the nucleotide[187,188] with the third glycine residue forming part of this loop. Substitution of the small side chain of glycine with the slightly larger threonine residue may disrupt the ATP-specific recognition, and confer some other substrate recognition. It may be worth reiterating at this point that a viral mutation of the third glycine residue to a lysine in *v-SRC* abolishes the transformation and catalytic activity of this oncogene.[189] It is also noteworthy, that this glycine interacts sterically with the conserved phenylalanine and glycine of the subdomain VII motif D-F-9 in the catalytic domain of the insulin receptor.[190] This conformation is involved in maintaining an open structure between the two lobes of the catalytic domain, and perhaps the altered glycine to threonine and phenylalanine to proline in JH2 suggests an alternate structural requirement.

Certain other subtle differences exist in the normally consistent spacing between key motifs in JH2 as compared with a PTK domain. For example, the spacing between both components of the ATP-binding site (I_a and II_a) is different for JAK1, JAK2, JAK3 and Tyk2 when compared with the broader protein kinase family. In JAK1 this spacing contains an extra seven amino acids, JAK2 and JAK3 an extra three amino acids, and Tyk2, an extra twenty-one amino acids (Fig. 10). Moreover, for JAK1, JAK2, Tyk2 and JAK3, the spacing between subdomains VI_a and VII_a in this region is also longer. Conversely, the distance between subdomains VII_a and IX_a in JAK1, Tyk2 and JAK3 is seven amino acids shorter that the corresponding region in the JH1 domain. It is worth noting that this subdomain in the PTK domain contains the putative autophosphorylation tyrosine residue, while in each JAK family member, this tyrosine is not present in the kinase-related/JH2 domain. The overall structure of this domain may be expected to be somewhat different from the catalytic domains of other members of the PTK and threonine/serine kinase families.

To date, no catalytic activity has been demonstrated for the JH2 domain, and it seems likely that the JH2 domain may allosterically regulate some other aspect of JAK function, perhaps even the activity of the PTK domain itself. A precedent exists for this type of regulatory role in the case of the atrial natriuretic peptide receptor (ANP receptor), wherein after ANP activation of this receptor, the C-terminal guanylate cyclase domain is regulated in an ATP-dependent fashion by a kinase-related domain reminiscent of that found in the JAK family.[191]

Other Functional Domains of the JAK PTKs

The seven homology domains defined from a simple comparison of the four mammalian members of the JAK family of PTKs (see above) might suggest a complex series of interactions with other proteins. At present, however, the function of most of these domains remains poorly understood. While most other nonreceptor protein tyrosine kinases possess one (often two) SH2 domain(s), the JAK family appears to be lacking this domain.[167] However, it appears that the highly conserved JH3 and JH4 domains, when taken together, bear a (rather tenuous) resemblance to an SH2 domain.[13,27] Comparison of the sequences of most of the known SH2 domains reveals the presence of a number of conserved subdomains. The availability of X-ray crystal structure information on several SH2 domains[30-35] has further underlined those elements which play important roles in the recognition of their cognate phosphotyrosine-bearing binding sites. Interestingly, close to the C-terminal boundary of JH4 lies the sequence -YVLRWS-, which has some degree of homology to the SH2 domain core sequence -FLVRES-, which plays an important role in the recognition of the phosphotyrosine residue itself. Additional subtle sequence similarities have been detected between SH2 consensus sequences and other portions of the JH3 and JH4 domains[192] (see Fig. 12). An alignment of these two domains from JAK1, JAK2 and Tyk2[166] reveals that while the similarity of the JH4 domain to the SH2 domains is most evident in the region surrounding the SH2 core sequence (FLVRES), the homology does not extend far in either direction beyond this region, and only reappears toward the C-terminal boundary. It may well be the case for the JAK family PTKs that this domain may represent a variation on the theme of SH2 domains, or an evolutionarily distant relative of a primordial SH2 domain. However, it may be that the lack of extensive homology suggests this homology is only a chance occurrence.

STRUCTURE/FUNCTION RELATIONSHIPS WITHIN THE JAK FAMILY OF PTKs

The multidomain structure of such molecules as S6 kinase[193] (possessing two serine/threonine kinase-related catalytic domains), the membrane ANP receptor-guanylate cyclase protein[191] (a kinase-related domain coupled to guanylate cyclase domain) and the yeast GCN2 protein[194] (a kinase-related domain coupled to a tRNA synthetase-related domain) are interesting echoes of the bidomain structure of the JAK family of PTKs. In particular, the unusual nature the kinase-related domain (JH2) of the JAK kinases is reminiscent of the JH2 domain of the atrial natriuretic peptide receptor (ANP-R), wherein, notwithstanding the overall structural relatedness of the kinase-related domain to the protein kinase family as a whole, there are significant deviations in detail between the precise amino acid sequences of certain key motifs and protein kinase consensus sequences.[4,5] In light of this, one may question the nature of the anticipated catalytic activity of this domain. One example has already been given above, in the conserved kinase motif subdomain VI_a motif, -VHG<u>N</u>VCTKNL- where all known protein

```
JAK1    QNGCHGPIC-EYAI.NKLRQEGSEEGMYVLRWSCT...DFDNILMTVTCFEKSEQVQGAQKQFKNFQIEVQKGRYSLHGSDRSFPSLGDLMSHLKKQILRTDNISFMLKRCCQPKP
JAK2    HSNCHGPISMDFAI.SKLKKAGNQTGLYVLRCSPK...DFNKYFLTFAVER.....ENVIEYKHCLITKNENGEYNLSGTKRNFSSLKDLLNCYQMETVRSDSIIFQFTKCCPPKP
TYK2    RDGIHGPLLEPFVQQAKLRP..LEDGLYLIHWSTS...HPYRLILTVAQRSQAPDGMQSLRLRKF.PIEQQDGAFVLEGWGRSFPSVRELGAALQGCLLRAGDDCFSLRRCCLPQP
         -GCHGPI---FAI -KLR--G-E-GLYVLRWS-----DF----LTVA---------------K---IE-Q-G-Y-L-G--RSFPSL-DL---LQ---LR-D-I-F-L-RCC-PKP
            || |    |    |       |·||||·|     ||                       .          |   |  | | || ||
          WYHGKI----A----L-------GSYLIRES---PGDFVLS--------------------R----------Y---G--R-F-SL-DL--YY---------------L-EPV
GAP-N     WYHGKLDRTIA.EERLR.QAGKSGSYLIRESDRRPGSFVLSFLSQT.NV........VNHFRI..IAMC.GDYY.IGG.RRFSSLSDLIGYYSHVSCLLKGE....KLLYPV
GAP-C     WFHGKISKQEA.YNLLM.TVGQACSFLVRPSDNTPGDYSLYF.RTSENIQ..R.......FKI.CPTPN.NQFM.MGG.RYYNSIGDIIDHYRKEQIVEGYY.....LKEPV
v-Crk     WYWGRLSRGDA.VSLLQ..GQRHGTFLVRDSGSIPGDFVLSV.SESSRVS......HYIVNSLGPAGGRRAGGE.[18]..FDSLPSLLEFYKIHYLDTT......TLIEPV
```

Fig. 12. Comparison of the JH3/JH4 domain region of JAK1, JAK2 and Tyk2 with the SH2 domains of GAP and v-Crk. The SH2 domains of GAP and v-Crk were compared with the JH3/JH4 regions of JAK1 (amino acids 425-536), JAK2 (amino acids 405-512) and Tyk2 (amino acids 449-555). Sequence similarity is shown by means of a line between the two sets of sequences. The similarity is greatest in the region of the FLVRES motif of the SH2 domain, however, there is little similarity with the C-terminal regions of the SH2 domains.

kinases have an invariant aspartic acid residue,[4,5] the JH2 domain possesses a conserved asparagine residue. While this divergence is particularly unusual, a similar variation is present in the ANP-R/guanylate cyclase kinase-related domain[191] and in the kinase of the receptor PTK *Erb-B3*,[195] further supporting the notion of an altered activity for the JH2 domain. This conserved aspartic acid residue has been shown to be essential for catalytic activity in the case of *v-fps*[196] and this is likely to be a general theme in the PTK family. Such a deviation from consensus suggests that other enzymatic properties, such as the capacity to phosphorylate substrates other than proteins, should be examined. Since the overall structural topology of this domain is fairly conserved, it is not unreasonable to postulate that it might act as a regulatory domain, perhaps to sequester ATP, or another substrate, without itself being catalytically active. Alternatively, a novel spectrum of amino acids may be recognized and phosphorylated by this unique domain. Whatever its function, the overall sequence and structure conservation of this domain in all JAK family members known to date, speaks to an important catalytic (or indeed structural) role for this domain in the signaling niche in which the JAKs function within the cell.

The important role of the PTK domain of JAK2 has been demonstrated in a number of cases. In one study, fusion of the PTK (JH1) domain of JAK2 to the intracellular region of the growth hormone receptor (GH-R) generated a chimeric receptor capable of triggering an appropriate transcriptional response in transfected cells,[197] while a fusion to the JH2 domain alone failed to signal appropriately. Inclusion of both the JH1 domain and the kinase-like domain (JH2) in this hybrid protein did not appear to affect the signaling from this chimera. Parallel studies in the granulocyte-macrophage colony stimulating factor receptor (GM-CSF-R)[198] system, have demonstrated the importance of the JH1 domain for signaling, and the further requirement of an intact N-terminal region of JAK2 was apparent in the case of GM-CSF-R,[198] although in these studies the JH2 domain appeared to be less important. Intriguingly, these observations are at odds with the data which has recently emerged from an analysis of IFNα/β signaling in a fibroblast cell line defective in its expression of Tyk2[199] (see below). Here, the importance of an intact kinase-like domain was demonstrated, since a mutant Tyk2 cDNA lacking the JH2 domain was unable to complement the IFNα/β pathway in this mutant cell line, whereas a mutant Tyk2 lacking the JH1 domain, was able to partially restore signaling. In all of these analyses, the mutation of each domain encompassed a relatively large segment of the JAK protein, and it remains possible that effects on the global structure of the JAK itself are appearing as functionally specific effects. Nonetheless, a second round of more subtle mutations will surely begin to bring this issue into focus in the near future.

EVOLUTION OF THE JAK FAMILY OF PTKs

The idea that the components of intracellular signal transduction pathways are essentially highly conserved throughout metazoa is an attractive one. This possibility has fueled the investigation of these signal transduction pathways in experimentally tractable organisms such as the fruit fly (*Drosophila melanogaster*), Zebrafish (*Danio rerio*) and the nematode worm (*Caenorhabditis elegans*). As already discussed, several non-mammalian JAK family members have been uncovered, suggesting that the structural format of the mammalian JAKs is one that evolved prior to the separation of the progenitors of the invertebrates from the progenitors of the vertebrates. If the comparative analysis described above is extended to other non-mammalian species such as F-JAK and *hopscotch*, it is clear that the overall architecture of the JH domains has been retained in all of the vertebrate JAKs, including F-JAK. However, the invertebrate JAK family member, *hopscotch*, has diverged substantially from the mammalian JAKs, specifically with respect to the more N-terminal domains. This may represent a genuine evolutionary drift, or might be due to the emergence of a different function for *hopscotch* in *Drosophila melanogaster*. In this respect it is worth noting that, to date, there are no cytokine-type receptors described in *Drosophila*, which may specifically preclude a role for *hopscotch* in the type of signal transduction pathway in which the mammalian JAKs now seem to operate.

With regard to the evolutionary conception of the dual kinase domain structure of the JAK family, there exists two scenarios, each with distinct but important ramifications, regarding the evolutionary development of a family of proteins such as the JAK family of PTKs. The extent of the JAK family across the phylogenetic tree is testament to a very early acquisition of the kinase-related domain by one of the following mechanisms, and it is distinctly possible that duplication and subsequent remodeling of the two kinase-related domains into the present form described above has occurred. Conversely, it is possible that the JH2 domain was "captured" by a protein already bearing a PTK-related domain, in the manner akin to that by which the SH2 domain of the nonreceptor PTKs has found its way into several other families of protein, such as *vav*,[21,22] *SRC*[200] and SHC.[201] Together, the mosaic nature of such multidomain proteins, including the PTK family, and the observation that these are functional elements (like the SH2 domain) shared between protein families, supports the notion of a "modular" approach to their construction. If the JAK family members have been constructed in such a manner, it is conceivable that the same JH2 domain may also be a component of other non-PTK proteins. Just as SH2 domains are present in PTKs and non-PTK proteins alike, it is a distinct possibility that a domain similar to the JH2 domain is present in other classes of proteins.

CHAPTER 4

Signal Transducers and Activators of Transcription: The STAT Proteins

The STATs are a family of latent cytoplasmic transcription factors. Upon activation by tyrosine phosphorylation, they homo- or heterodimerize and translocate to the nucleus where they interact with specific DNA sequences upstream of target genes (STAT-binding elements or SBEs) (Fig. 24; also see chapter 5 for a more detailed description). At present, six mammalian members of the STAT family are known, namely, STAT1,[202] STAT2,[203] STAT3[204,205] also known as acute phase response factor (APRF),[206] STAT4,[204] STAT5 (also known as mammary gland factor [MGF]),[207] and IL-4-STAT[208] (also known as STAT-6).

The STAT family was first identified during an investigation into the regulation of gene transcription by IFNα/β.[202,203,209-211] In early work, a group of IFNα/β inducible genes were characterized and the genes encoding them analyzed.[212-214] From these studies, a single DNA response element, the interferon-stimulated response element (ISRE) (see chapter 5), was uncovered and subsequently demonstrated to confer inducibility by IFN.[215] Furthermore, an ISRE was shown to be a feature held in common by many of the IFNα/β inducible genes examined, as well as some IFN inducible genes. Intriguingly, the transcriptional induction of genes such as

GBP[216] involved the activation of apparently cytoplasmically localized DNA-binding complexes/transcription factors. Although GBP induction was mediated by different DNA-binding complexes specific for each type of IFN, these binding complexes did possess a common protein (p91) (now known as STAT1[217]), giving rise to the notion of a new class of transcriptional activators that might be utilized by a range of individual signaling pathways. Indeed, this turned out to be the case, with the prototype of these ligand-dependent activators of transcription, the interferon-stimulated gene factor 3 (ISGF3) which was purified and characterized.[203,211,218,219]

Each of the protein components of ISGF3 were subsequently purified, sequenced and cloned,[202,211,218] and it became apparent that three of these proteins (p84, p91 and p113) were highly related to each other. The fourth component, p48, turned out to be a DNA binding protein with a *myb* homology domain.[220] The three new family members were renamed the signal transducers and activators of transcription (STATs),[217,221] and have become the archetypal members of a new family of related transcription factors, which at present number six. The role(s) that the STATs play in signal transduction downstream of a large number of cytokines is now beginning to be appreciated, and in combination with the JAK family of PTKs, these proteins are an important (and indeed eponymous) component of the JAK-STAT pathway.

STAT1 (p91/p84)

The cDNA for STAT1 was initially isolated by screening a human cDNA library (16 hour IFNγ- and 45 min IFNα-treated HeLa cells) with degenerate oligonucleotides whose design was based upon amino acid micro-sequence information from ISGF3 purified from IFN-induced HeLa cells.[202] Analysis of the cDNA sequences revealed that p91 and p84 (now designated STAT1α and STAT1β, respectively) arose from two alternately processed RNA products of the same gene.[202] The STAT1α coding sequence of 2217 bp predicted a protein product of 86 kDa. While 86 kDa was somewhat shorter than the 91 kDa protein that was purified ex vivo, in vitro coupled transcription/translation of the STAT1α cDNA confirmed the smaller size, leading the authors to suggest this difference may be due to post-translational modifications.[202] STAT1β arises from an alternate splicing variant that appears to diverge from the sequence encoding STAT1α at nucleotide 2286, beyond which point the sequences are completely different. Northern blot analysis demonstrated this difference clearly, with probes to common regions of both forms indicating two mRNA transcripts (3.1 kb and 4.4 kb), while probes against the unique 3'-ends of STAT1α and STAT1β hybridized singly to the 4.4 kb and 3.1 kb transcripts, respectively.[202] Thus, the proteins encoded by STAT1α and STAT1β mRNAs are identical up to the conserved Y701, however, beyond this element the sequences of STAT1α and STAT1β diverge completely. Both isoforms possess domains which are related to SH2 and SH3-like domains.[202]

STAT2 (p113)

STAT2 cDNA was isolated from a human cDNA library using a degenerate oligonucleotide, the sequence of which was determined from micro-sequence protein data from the purified p113 component of ISGF3.[211] The predicted protein is 851 amino acids (2553 bp), and Northern blot analysis revealed a single mRNA species of 4.8 kb.[211]

STAT3 (Acute Phase Response Factor-APRF)

Acute phase response factor (APRF), is a transcription factor complex capable of binding to IL-6-responsive elements present in the promoters of plasma proteins that are regulated in response to inflammation (known as acute-phase proteins). This transcriptional activator, now known to be composed of homodimers of STAT3, was identified and cloned contemporaneously by a number of groups,[204,206] and is anti-genically related to STAT1.[222]

The mouse STAT3 cDNA encodes a polypeptide of 770 amino acids with a predicted molecular weight of 88 kDa.[204,206] Northern blot analysis revealed a single mRNA of 4.8 kb for both human and mouse STAT3, with transcripts being present in all human[206] and mouse[204,206] tissues tested. Furthermore, stimulation of mouse liver with IL-6 for 3 hours induced the level of STAT3 mRNA 5-fold,[206] confirming the identity of the cloned STAT3 as an IL-6-inducible factor.

Two isoforms of STAT3 have been uncovered, and are the products of differential splicing of transcripts from the same gene. These variants are known as STAT3α and STAT3β, respectively, and appear to have different functions in determining the specificity of promoter activation.[223]

STAT4

The identification of STAT4 came about from low stringency cDNA library screening[204] and from PCR screens using degenerate STAT specific primers.[224] The open reading frame of mouse STAT4 encodes a protein of 749 amino acids, of approximately 89 kDa.[204] An mRNA transcript of 3.3 kb was detected by Northern blot analysis in a limited number of tissues.[224] In the myeloid cells, which express the highest levels of STAT4, a larger 3.6 kb transcript was detected in conjunction with the 3.3 kb transcript.[224] STAT4 expression is largely restricted to early myeloid precursors, and appears to be down regulated during development of erythroid lineages.[224] Interestingly, the only other site of significant expression of STAT4 is the developing spermatogonia of the testis. At present the significance of this observation is unknown.

STAT5 (MAMMARY GLAND FACTOR-MGF)

Prolactin, like the interferons, triggers the activation of cytoplasmic transcription factors capable of binding to the enhancer of the IRF-1 gene (a transcription factor known as PRL-induced factor [PRLIF] and/or PRL-stimulated factor [PRL-SF]),[225,226] and to a regulatory element in the β-casein gene (a transcription factor complex known as mammary gland factor [MGF]).[207,227] Of these transcription factors, MGF has now been purified and shown to be composed of homodimers of a single protein. A cDNA encoding this protein was subsequently cloned, and the protein has been renamed STAT5 on the basis of homology with other STAT proteins.[207,227] STAT5 was originally identified in mammary epithelial cells of lactating animals,[207] and displays all the characteristic activation criteria that are common to members of the STAT family.[227]

In later studies, an IL-3-dependent cell line (MC/9) cDNA library was screened with the ovine STAT5 (MGF) cDNA[228] and two different but highly related clones were isolated. These isolates were designated STAT5A and STAT5B. At the nucleotide level the identity was 80% between the ovine STAT5 and both murine STAT5A and STAT5B, while the amino acid comparison was greater than 90%. Transcripts encoding each variant were present in a wide range of tissues—brain (only STAT5B), heart, lung and kidney, but not liver—and cell types—T cells, B cells, macrophages and myeloid cells.[228] The STAT5B coding sequence is 7 amino acids shorter than STAT5A[228,229] and results from the use of a different splice site.

STAT6 (IL-4 STAT)

The IL-4-induced DNA-binding factor, termed STF-IL-4[230] or IL-4 STAT[208] was purified and the mRNA encoding its constituent polypeptide was molecularly cloned from human sources.[208] The predicted amino acid sequence of this protein displayed a high degree of similarity with the prolactin-induced DNA-binding factor (MGF or STAT5[207]) and it was named IL-4 STAT. The designation of this factor has been altered to STAT6, in keeping with the nomenclature of the remainder of the STAT family. STAT6 has an apparent molecular weight (by SDS-PAGE analysis) of approximately 100kDa.[208] The human STAT6 coding sequence predicts an open reading frame of 2544 bp (848 amino acids) corresponding to a calculated molecular weight of 94 kDa.[208,231] Mouse STAT6 was isolated from a MEL cell library using a human cDNA probe,[231] derived from sequences obtained by screening databases of human genes identified by the human expressed sequence *"tag"* (EST) method.[232-235]

GENES ENCODING THE STATs

The chromosomal location of each of the mouse STAT genes has been carried out[224] and an interesting clustering pattern has been uncovered (see Table 3). Thus, STAT1 and STAT4 are located close together on mouse chromosome 1 (at a region syntenic with human chromosome 2q32-q34), while STAT2 and STAT6 are similarly colocalized to mouse chromosome 10 (in a region of synteny with human chromosome 12q13-q14.1). Finally, STAT3 and STAT5 colocalize to mouse chromosome 11 (syntenic with human chromosome 17q11.2-q22). Taken together, these data suggest that the promordial mammalian STAT gene has been duplicated once, and the duplicated unit replicated twice more. While the hunt for new STATs in the mammalian genome is continuing, it seems that the assignment of the chromosomal locations of the known STATs into three neat blocks of two could mean that the entire complement of STATs may already be known.

The genomic organization of STAT2 and a large portion of STAT1 have been determined.[236] Both genes are complex, with the STAT2 protein being encoded by 24 distinct exons, some encoding stretches of the protein as small as seven amino acids. The location of all of the intron/exon boundaries for STAT1 and STAT2 are coincident, reconfirming their common ancestry. If this complexity of organization extends to the other STAT genes then there is obviously considerable scope for differential splicing within the STAT family.

CONSERVED FEATURES

The STATs possess seven homology domains held in common by each of the six members of the family (Fig. 13). A further two elements, potential sites of phosphorylation, are strongly conserved within the family. At present, five of these domains have been clearly implicated in STAT function:

(1) the N-terminal 140 amino acids;[237]
(2) a leucine heptad repeat region located approximately 200 amino acids from the N-terminus in STAT1;[237]
(3) the SH2 domain;[238,239]
(4) a tyrosine residue C-terminal of the SH2 domain,[238] the prototype of which is located at position 701 of human STAT1 (Y701)[217,238,240]; and
(5) a serine phosphorylation site located at position 727 in STAT1.[241]

The SH3 domain which is located immediately N-terminal to the SH2 domain has not yet been ascribed a function.

THE N-TERMINAL DOMAIN

The N-terminal 50 amino acids of each of the six known mammalian STATs are quite well conserved (Fig. 14), exhibiting

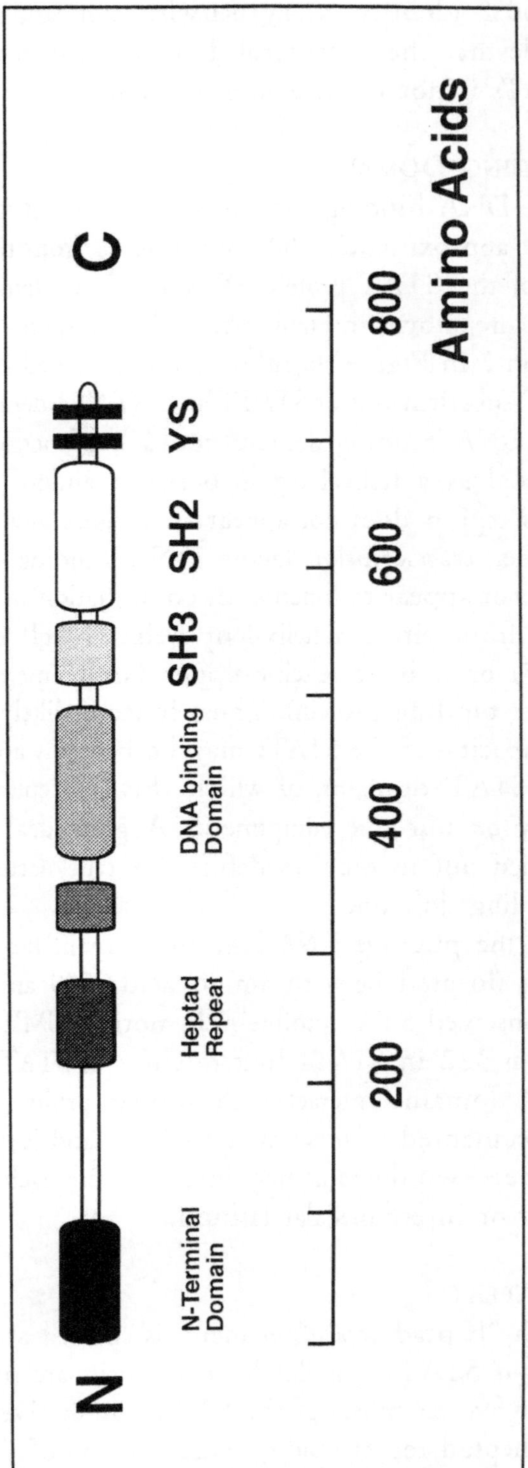

Fig. 13. Domain structure of the STATs. Comparison of the sequences of the six known mammalian STATs revealed six regions of high conservation. The N-terminal domain, the heptad repeat, the DNA binding domain, the SH3 domain, the SH2 domain and the sites of serine (S) and tyrosine (Y) phosphorylation are shown. These domains are discussed in depth in the text.

approximately 50% identity in any pairwise comparison of two STATs. The role that the N-terminal domain plays in the function of the STATs is not understood at present.

The DNA Binding Domain

The putative DNA binding domain of members of the STAT family is located approximately 400 amino acids from the N-terminus of each of the STAT proteins.[242] The STAT family DNA binding domains are shown in Figure 15, and correspond to STAT homology domain 2 in Figure 16. Information obtained from both "mix and match" substitutions of STAT1 and STAT3 demonstrated that the STAT DNA binding domain could be functionally defined to be located in a helical region between amino acids 400 and 500.[243] This region does not appear to possess any sequence similarity to other transcription factor DNA binding domains. Moreover, it did not appear to function in combination with dimerization domains from either a helix-loop-helix (HLH) transcription factor (E47) or a basic leucine-zipper family member (the CCAAT/enhancer binding protein). Thus, it seems likely that the DNA binding capacity of the STATs may be brought about by a combination of STAT domains, of which this particular homology domain may be but one component. A great deal of work needs to be carried out in order to define the true nature of the STAT DNA binding domain.

Interestingly, the putative DNA binding domain lies between the SH3 domain (located between amino acids 500 and 580 in STAT1) and a conserved STAT proline-rich motif (PCMPXXPXXP located at position 322 in STAT1 [but missing in STAT5]). The notion that SH3 domains interact with similar proline-rich elements is well documented in other systems,[244,245] and it remains a possibility that these two domains may interact with each other in an intramolecular or intermolecular fashion.

The Heptad Repeat

A leucine-rich "heptad repeat" element is located at position 200-220 in human STAT1. Similar helical motifs are present in each of the other five members of the STAT family. Deletion of this leucine-rich heptad repeat results in inactivation of STAT1.[237]

```
STAT1     MSQWFELQQLDSKFLEQVHQLYDDSF-PMEIRQYLAQWLEKQDWEHAAYVD    50
STAT2     MAQWEMLQNLDSPFQDQLHQLYSHSLLPVDIRQYLAVWIEDQNWQEAALGS    51
STAT3     MAQWNQLQQLDTRYLKQLHQLYSDTF-PMELRQFLAPWIESQDWAYAASKE    50
STAT4     MSQWNQVQQLEIKFLEQVDQFYDDNF-PMEIRHLLAQWIETQDWEVASNNE    50
STAT5     MAGWIQAQQLQGDALRQMQVLYGQHF-PIEVRHYLAQWIESQPWDAIDLDN    50
STAT6     MSLWGLVSKMPPE---KVQRLYVD-F-PQHLRHLLGDWLESQPWEFLVGSD    46
```

Fig. 14. The N-terminal domains of the six mammalian STATs.

```
STAT1     TNGSLAAELRHLQLKEQK-NAGNRTN-EGPLIVTEELHSLSFETQLCQPG--LVIDLETT     451
STAT2     QSQGLIWDFGYLTLVEQRSGGSGKGSNKGPLGVTEELHIISFTVKYTYQG--LKQELKTD     449
STAT3     NNGSLSAEFKHLTLREQRCGNGGRANCDASLIVTEELHLITFETEVYHQG--LKIDLETH     457
STAT4     SNGSLSVEFRHLQPKEMKCSTGSKGN-EGCHMVTEELHSITFETQICLYG--LTINLETS     448
STAT5     ATGTLSAHFRNMSLKRIK-----RADRRGAESVTEEKFTVLFESQFSVGSNELVFQVKTL     461
STAT6     PGNCCSALFKNLLLKKIK-----RCERKGTESVTEEKCAVLFSASFTLGPGKLPIQLQAL     406

STAT1     SLPVVVISNVSQLPSGWASILWYNMLVTEPRNLSFFLNPPCAWWSQLSEVLSWQFSSVTK     511
STAT2     TLPVVIISNMNQLSIAWASVLWFNLLSPNLQNQQFFSNPPKAPWSLLGPALSWQFSSYVG     509
STAT3     SLPVVVISNICQMPNAWASILWYNMLTNNPKNVNFFTKPPIGTWDQVAEVLSWQFSSTTK     517
STAT4     SLPVVMISNVSQLPNAWASIIWYNVSTNDSQNLVFFNNPPSVTLGQLLEVMSWQFSSYVG     508
STAT5     SLPVVVIVHGSQDHNATATVLWDNAFAEPGRVP--FAVPDKVLWPQLCEALNMKFKAEVQ     522
STAT6     SLPLVVIVHGNQDNNAKATILWDNAFSEMDRVP--FVVAERVPWEKMCETLNLKFMAEVG     463
```

Fig. 15. The DNA-binding domains of the six mammalian STATs. Amino acid positions of the most C-terminal residue are shown on the left of each line.

```
STAT1     DTILELIKNDLLCLWNDGCIMGFISKERERALLKDQQPGTFLLRFSESSREGAITFTWVE     618
STAT2     DKILELVHDHLKDLWNDGRIMGFVSRSQERRLLKKTMSGTFLLRFSESS-EGGITCSWVE     616
STAT3     DNIIDLVKKYILALWNEGYIMGFISKERERAILSTKPPGTFLLRFSESSKEGGVTFTWVE     625
STAT4     EAILDLIKKHILPLWIDGYIMGFVSKEKERLLLKDKMPGTFLLRFSES-HLGGITFTWVD     614
STAT5     DGVMEVLKKHHKPHWNDGAILGFVNKQQAHDLLINKPDGTFLLRFSDSE-IGGITIAWKF     633
STAT6     DGVLDLTKRCLRSYWSDRLIIGFISKQYVTSLLLNEPDGTFLLRFSDSE-IGGITIAHVI     577

                    --βA--      --αA--      --βB(FLVRES Motif)--

STAT1     RSQNGGEPDFHAVEPYTKKELSAVTFPDIIRNYKVMAAENIPENPLKYLYPNIDKDHAFGKYYS   682
STAT2     H-QDDDKVLIYSVQPYTKEVLQSLPLTEIIRHYQLLTEENIPENPLRFLYPRIPRDEAFGCYYQ   679
STAT3     K-DISGKTQIQSVEPYTKQQLNNMSFAEIIMGYKIMDATNILVSPLVYLYPDIPKEEAFGKYCR   688
STAT4     QSE-NGEVRFHSVEPYNKGRLSALAFADILRDYKVIMAENIPENPLKYLYPDIPKDKAFGKHYS   677
STAT5     DSPE---RNLWNLKPFTTRDFSIRSLADRLGDL----------SYLIYVFPDRPKDEVFSKYYT   684
STAT6     RGQD-GSPQIENIQPFSAKDLSIRSLGDRIRDL----------AQLKNLYPKKPKDEAFRSHYK   630
```

Fig. 16. The SH2 domains of the six mammalian STATs. The location of two of the β-sheets and one of the α-helices of a typical SH2 domain are shown. Amino acid positions of the most C-terminal residue are shown on the left of each line.

In this mutant, STAT1 fails to become phosphorylated upon tyrosine in response to IFNα/β, resulting in the loss of function of this protein. Since this part of the protein is some distance from the tyrosine phosphorylation site located at position 701 in STAT1 (at least in terms of the linear structure of STAT1), it may be that this domain is playing a role in the regulatory process which results in phosphorylation of this site, rather than defining the site per se. It seems likely that the structure of STAT1 may be particularly complex. These data suggest that there are a number of interactive domains required for its cellular function.

The SH2 Domain

It has come to be widely accepted that SH2 domains possess a high degree of sequence specificity with respect to their interactions with phosphotyrosine (PY) residues in proteins.[13,111] The detection of significant homology between other SH2 domains and the region lying between amino acids 590 and 690 in human STAT1, suggested that this protein, and its five relatives, are members of a unique class of SH2 containing transcription factors. The SH2 domains of the STATs are among the most highly conserved regions of this family, exhibiting over 50% identity in a pairwise comparison between any two STATs. Central to the notion that this domain is indeed an SH2 domain is the presence of a highly conserved "FLVRES" motif[201] located close to the N-terminal boundary of this domain. Mutation of the arginine residue in this element of the STAT1 SH2 domain rendered STAT1 incapable of contributing to an active ISGF3 complex,[237] thus adding weight to the notion that this homology region was indeed a bona fide SH2 domain.

The crystal structures of a number of SH2 domains are now available, and it appears that the FLVRES motif is responsible for the docking of the phosphate upon the tyrosine residue itself. Importantly, the specificity of the PY/SH2 domain interaction appears to reside within elements in the C-terminal portion of the domain. In the case of the STAT family, the N-terminal portions of the STAT SH2 domains (particularly around the FLVRES element) are especially well conserved, while the C-terminal sequences are somewhat less well preserved. This may indicate the likelihood of a difference in phosphotyrosine binding specificity.[206] This notion is further explored in chapter 7.

It has become abundantly clear that this domain has an important role to play in the function of the STATs, and indeed appears to be the key to the specificity exhibited by each of them in eliciting the generation of an appropriate homo- or heterodimeric STAT transcription factor complex (or "STF"; see next chapter), as well as being the "glue" which holds the STAT complexes together. Thus, the direct interaction of the SH2 domain of one STAT1 molecule and the Y701 phosphotyrosine residue on a second STAT1 protein has elegantly demonstrated the likely function of each of these domains in the coordination of STAT dependent signaling.[238] Similarly, the affinity of the STAT1 SH2 domain for a tyrosine residue located at position 440 (Y440) in the cytoplasmic tail of the human IFNγ-R,[246] lead to the suggestion that the SH2 domain of STAT1 might also serve to coordinate the interaction of STAT1 with the IFNγ-R. After tyrosine phosphorylation (most likely by a receptor associated JAK), the phosphorylated STAT then homodimerizes[238] (or perhaps heterodimerizes with another STAT) and becomes able to migrate to the nucleus to initiate transcription events. The precise choreography of these SH2 phosphotyrosine binding events are outlined in more detail in chapter 7.

The "Y701" Phosphorylation Site

It is now well established that the major mechanism of STAT activation is through tyrosine (and indeed, serine) phosphorylation.[217,219,237,241,246-250] The activation of STAT1 within ISGF3 and GAF (γ-activated factor) following IFNα and IFNγ stimulation respectively, required phosphorylation of a tyrosine residue located within the C-terminal tail, immediately adjacent to the boundary of the SH2 domain (Tyr 701 in STAT1).[251] The phosphorylation of this site appeared to result in the homodimerization of STAT1 via an intermolecular SH2-phosphotyrosine interaction,[238] thereby producing an active GAF. Further support of this phosphotyrosine being a binding site for an SH2 domain on a partnering STAT protein was that mutation of the arginine 602 (the critically conserved arginine residue within the SH2-indicative FLVRES motif[15]) resulted in the abrogation of activation of STAT1 and hence, prevention of dimerization.[238,250] The notion of an SH2/PY interaction involving the tyrosine at 701 in STAT1 and the SH2 domain of a partnering STAT was thus born.[238]

Soon after the identification of tyrosine 701 as the phosphorylation site required for activation of STAT1, other studies demonstrated that the homologously located tyrosine residue in STAT2 (tyrosine 690) was also a phosphorylation target.[237] These data further demonstrated that the phosphorylation of tyrosine 690 occurred in the absence of STAT1.[237] The ordered assembly of the ISGF3 complex at the IFNα/β receptor is discussed later, in chapter 7.

A tyrosine phosphorylation site for STAT3 is yet to be determined empirically, however, a strong candidate site can be identified by examination of the amino acid sequence surrounding the tyrosine 705, a tyrosine located in an homologous position to tyrosine 701 of STAT1. This speculation appears to have some foundation, and the laboratory of J.E. Darnell Jr. has reported unpublished data[242] suggesting that mutation of tyrosine 705 of STAT3 to phenylalanine blocks phosphorylation of STAT3.

STAT4 was the last STAT protein to be assigned a role in cytokine signal transduction, and has now been shown to be activated in response to IL-12.[252,253] IL-12 induces, amongst others, a factor containing both STAT4 and STAT3.[252] Due to the restricted activation of STAT4, no in-depth analysis of the phosphorylation sites of this STAT has been published to date, however, unpublished results (referred to in ref. 231) suggest that the phosphorylation site of STAT4 has been located at position 694 in the mouse STAT4 sequence.

Examination of the sequence in the C-terminal tail of STAT5 revealed the presence of three tyrosine residues that lie within similar sequence contexts to that seen for the STAT1 tyrosine 701. Mutagenesis of each of these tyrosines determined that Tyr 694 (in the analogous position relative to the C-terminal boundary of the SH2 domain as STAT1) was phosphorylated in response to

PRL treatment, supporting the idea that the same mechanism applies to the assembly of other STAT dimers.

STAT6 (IL-4 STAT) homodimerizes in response to IL-4, as demonstrated by chemical crosslinking studies.[208] This study also ascertained the likely phosphotyrosine binding sites on the IL-4R chain by the use of the inhibitory phosphopeptides Y578 and Y606 from the IL-4R chain, to disrupt IL-4 STAT homodimerization.[208] The location of the STAT6 C-terminal phosphorylation site thus remains to be confirmed. However, it seems likely that the tyrosine located at position 641 (RGY*VPA) is the most likely candidate, based upon a sequence comparison of the C-terminal regions of each of the STATs.

The "S727" Phosphorylation Site

Serine phosphorylation has only recently been shown to play a role in the maximal activation of STAT1 and STAT3.[241,249,254] In studies examining the cellular responses in the differentiation of the promonocytic cell line U937 into monocytes, it was noted that the overall level of STAT1 phosphorylation increased in response to IFNγ, but that this induced phosphorylation was not on tyrosine but on serine.[249] The major serine phosphorylation target on STAT1 in vivo was mapped to serine 727.[241] Intriguingly, this site matches with the known MAP kinase (mitogen activated protein kinase) recognition sequence,[255] and indeed, a peptide containing this sequence was a demonstrable target for MAP kinase in vitro.[241] This site, however, has not yet been demonstrated as an in vivo target. Notwithstanding, it is clear from these studies that serine phosphorylation is required to achieve maximal STAT activation at the level of DNA binding. It is worth noting that the conserved MAP kinase consensus sequence is not present in the splice variant STAT1β,[202,236] nor is it present in STAT2[211] or STAT6.[208]

The most extensive phosphorylation studies on STAT3 have arisen from the perspective of serine phosphorylation. Observations that STAT3 experienced a time- and ligand-dependent retardation following induction by several ligands[254] and that the serine kinase inhibitor H7 abolished this mobility difference, prompted the investigation of the involvement of serine phosphorylation in STAT3 activation. In response to IL-6, STAT1 and STAT3 form homo- and heterodimers capable of binding an element in the c-Fos promoter.[205,256-258] The stability of the STAT3 homodimers (and interestingly, not the STAT1-STAT3 heterodimers) is compromised by the inhibition of serine phosphorylation of STAT3 by the use of the serine-specific phosphatase PP2A.[254] Furthermore, STAT3 also possesses a consensus MAP kinase target sequence (at serine 727), homologous to that present in STAT1.[255] Mutation of serine 727, indeed abolishes the formation of the slower migrating form of STAT3 (presumed to be serine phosphorylated in addition to tyrosine phosphorylated).[241]

Finally, the requirement for serine phosphorylation of STAT3 homodimers appears to correlate inversely with the affinity of the complex for a given response element; such that the high affinity binding to an SIE (*sis*-inducible element) is relatively unaffected, unlike the lower affinity binding to the site within the Ly-6E gene promoter, which is abolished in the presence of serine-specific phosphatases.[254]

The SH3 Domain

The SH3 domain is located between amino acid residues 500 and 580 in human STAT1. Studies in other systems have shown that domains such as these are able to bind proline rich elements in proteins, or are capable of modulating the affinity and specificity of SH2 domains. However, apart from the highly speculative notion that the SH3 domain is involved in an as yet hypothetical interaction with the proline-rich motif located at position 322 in STAT1, no function has been ascribed to this domain.

GENERATION OF STRUCTURAL COMPLEXITY WITHIN THE STAT FAMILY

While the alternative mRNA splicing of STAT1, STAT3 and STAT5 genes to produce structurally (and functionally) distinct members of the family appears to be a common theme in STAT biochemistry, there are differences in the location of the differential splice points within the individual STAT genes. This suggests that these variants have arisen as independent mechanisms of generating diversity in the gene family, or predicts the existence of a range of other splice variants of each of these genes which have yet to be discovered. As noted above, the complexity of the genes for STAT1 and STAT2[236] lends itself in support of this latter notion. This degree of complexity of isotype production is further compounded by a recently discovered propensity for post-translational modification within the STAT family.[241] As described above, both tyrosine and serine phosphorylation of STATs have been observed to play an important role in their function. These two mechanisms, coupled with the ability of individual STATs to form transcriptional complexes by both homodimerization and heterodimerization, leads ineluctably to the conclusion that a relatively small family of six STATs can generate an immense range of transcription factor complexes, each subtly distinct from the other.

SUMMARY

The mammalian STATs are a family of (at present) six structurally related proteins. They each possess a range of functional domains which serve particular aspects of their cellular function. The role of these proteins in the intracellular signal transduction pathway known as the JAK-STAT pathway is described in detail in chapter 7, while further details which do not yet fit into our current notions of the biochemistry of these molecules can be found in chapter 9. Suffice it to say, the STAT proteins were discovered as components of transcriptional complexes. In order to understand their role in the JAK-STAT pathway we are obliged to examine the DNA elements through which they regulate the transcription of their target genes. These elements, known as "STAT-Binding Elements," are the subject of the next chapter.

CHAPTER 5

STFs: STAT-Containing Transcription Factors

The term signal transducing factor (STF) was coined recently[257,259] to draw together an apparently diverse group of transcription factor complexes which are elicited following stimulation of cytokine receptors and PTK receptors.[257] These STFs were originally revealed by dint of their ability to bind to specific DNA elements (known as STAT-binding elements or SBEs; see below), as demonstrated by electrophoretic mobility shift assays (EMSAs). Although ach distinct cytokine evoked a particular STF, the composition of all of these STFs possessed a common theme, including the reactivity of some of the protein components with antiphosphotyrosine antibodies and the presence of STAT1 or a cross-reacting protein. Given the likely involvement of members of the STAT family of transcription factors in the cytokine-induced generation of transcription factor complexes such as these, it seems plausible that STFs are a functional manifestation of specific STAT-containing transcription complexes.

The issue of specificity of cytokine signal transduction hinges upon the ability of particular receptors to recruit a specific constellation of STFs. These in turn undertake the task of activating the transcription of those genes which define the cellular response to that particular cytokine. This process is an immensely complex one, relying as it does upon both subtle and gross differences in the affinity of each STF for a particular INFγ-activated (GAS)-like element. In this chapter we have attempted to address the underlying themes of how STAT complexes bind to specific DNA elements, and while this chapter is not exhaustive, most of the important notions have been addressed.

DNA ELEMENTS INVOLVED IN STF-MEDIATED ACTIVATION OF GENE EXPRESSION

Historically, there were two main subclasses of DNA response element found in the promoters of genes activated downstream of the JAK-STAT pathway. These are: the interferon-stimulated response element (ISRE)[260], an element located upstream of IFNα/β inducible genes; and the GAS[216]-like sequences located upstream of IFNγ-inducible genes. As will become clear these two classes of STAT binding elements are variations on the same basic motif. The GAS element was first identified as the ligand of an IFNγ-inducible STAT1-containing STF complex, known as gamma-activated factor (GAF).[217] Since then, GAS-like elements which bind STFs induced by other cytokines have been shown to consist of the same basic DNA motif with subtle sequence differences in individual cases found in specific genes. GAS-like elements are ligands for a wide range of STAT-containing complexes (not necessarily comprising STAT1). ISREs, on the other hand, appear to possess the ability to bind only STAT2-containing complexes, but vary in sequence composition between genes[260] such that the IFN-inducible genes 6-16 and ISG-15, for example, are both activated by IFNα, but have different core ISRE sequences.[261,262]

ESSENTIAL FEATURES OF STAT-BINDING ELEMENTS

There are three essential features of STAT-binding elements (SBEs) (Table 4):

(1) The core motif, comprised of the sequence

$$TT\text{-}(X)_n\text{-}AA$$

Where X is a C/G rich, primarily palindromic, sequence of length n.

(2) The core spacing between the palindromic A/T residues. For $(X)_n$, n can be 2, 3, 4, 5 or 6. This is the main determinant of the specificity of STAT binding; and

(3) The composition of the sequence between the palindromic A/T residues.

Table 5 shows the sequences of a range of natural SBEs found in cytokine responsive genes, alongside the type of STF which they bind. An obvious pattern emerges from which some of the finer points of specific DNA binding of STATs can be deduced. Firstly, there is a preference for a particular core spacing for each class of cytokine responsive genes. For example, classical GAS elements uniformly possess a five base pair core spacing, whereas IL-6 responsive genes possess a core spacing of four base pairs. For IL-4 responsive genes a core spacing of six base pairs appears to be preferred.

While these data were in themselves compelling, the notion that the cytokine-specific core spacings were, in turn, the basis of the specificity of response required further analysis. Several interesting studies have been undertaken in this area.[263,264] In one important study,[263] an artificial SBE was designed with the sequence:

$AACC(T)_nGGAA$ -Oligo series SpA

Individual variants of this sequence were generated where n varied from 0 to 3, and the ability of STFs generated by treatment of cells with a range of cytokines (IFNγ, IL-6, GM-CSF, IL-4) were tested for by EMSA for binding to each of these SBEs. The core spacing preferences exhibited by these cytokine induced STFs were reconfirmed in this experiment. Thus when n = 0 (i.e., a four base pair spacing lies between the palindromic AA-TT) there was a preference for IL-6 inducible STAT3 containing STFs, whereas when n = 2 (i.e., a six base pair spacing lies between the palindromic AA-TT) a preference for IL-4 inducible STAT6 containing STFs was observed. A second series of experiments was then performed, where the sequence composition of the spacer between the AA-TT palindrome was varied. Thus, using the artificial SBEs

$AACC(C)_nGGAA$ Oligo series SpB
and $AAC(C)_nGGAA$ Oligo series SpC

the same range of cytokine induced STFs were assayed for binding. In this series of experiments some of the subtleties of STAT

Table 4. ISRE elements

ISRE Elements	Sequence
ISRE from ISG15	GATCCATGCCTCGGGAAA**GGGAAA**CCGA**AACT**GAAGCC
ISRE core from ISG6-16	**GGGAAA**ATGA**AACT**
ISRE core consensus	**GGGAAA**(N)G**AACT**

The sequences of a number of known ISRE elements are shown aligned with each other; a consensus sequence is shown below.

Table 5. GAS elements

GAS-like Elements	Sequence
Human c-fos SIE	CAG**TTC**CCGTC**AA**TC
High affinity SIE (M67)	CAT**TTC**CCGT**AA**ATC
Ly-6E GAS	ATA**TTC**CTGT**AA**GTG
IRF-1 GAS	**TTC**CCCG**AA**
GBP GAS	**TTC**CCAG**AA**
pim-1 GAS	ACAACATCCC**TTC**CCAG**AA**ATCAGGATTC
β-casein PRE*	AGAT**TTC**TAGG**AA**TTCAATCC
APRE α_2-macroglobulin (IL-6RE)	CCA**TTC**CCGT**AA**GAG
FcγRI	GTAT**TTC**CCAG**AA**AAGGAAC
Iε	TCAAC**TTC**CAAG**AA**CAGAA
Consensus GAS-like	**TTC**NNNN**AA**

The sequences of a number of known GAS elements, and the genes from which they come, are shown aligned with each other; a consensus sequence is shown below.

binding were uncovered. For example, using oligo series SpC, when n = 2 (i.e., a four base pair spacing lies between the palindromic AA-TT) both STAT3 homodimers and STAT1/STAT3 heterodimers produced binding complexes. However, when n = 3 (i.e., a five base pair spacing lies between the palindromic AA-TT) STAT3 homodimers failed to bind and STAT1/STAT3 heterodimers bound only very weakly, whereas STAT1 homodimers bound well. Finally, it was observed that the GAS-like elements

	$TTC(T)_nGAA$	Oligo series SpD
and	$TTCT(C)_nAGAA$	Oligo series SpE

failed to bind STAT1 and STAT3 complexes regardless of spacing.

Taken together, these experiments on natural and artificial SBEs provide a basis for the binding specificity of STAT-containing complexes (STFs). The SBEs for most cytokine induced STAT-containing complexes can thus be fitted into this general framework. What of the ISRE? The core element of the ISRE is

G/AGGAAN(N)GAACT

and this cannot easily be fit into the same basic framework as the GAS-like SBEs. It is noteworthy that the IFNα/β inducible STF, ISGF3 is composed of at least one other protein (p48) in addition to STAT1 and STAT2, and that this protein also contacts the DNA to which the complex binds.[265] Perhaps the ISRE represents a hybrid site, composed of elements similar to the GAS-like sequences described above, coupled with additional specific elements which are defined by the p48 component of the complex.

Mutagenesis and DNA-binding studies have shed some light on this question.[265] The authors of this study propose that the TTT sequence present in the ISRE that STAT1 makes contact with, is only a half-site (of a GAS element) for this complex, and that there is further stabilization across the remainder of the element with either STAT2 or p48. Thus, the notion of a half-site, and composite elements made up of half-sites with different specificities is an intriguing one, and leads naturally to the generation of the fine-tuned specificity required of signal transduction processes.

STAT-CONTAINING SIGNAL TRANSDUCTION COMPLEXES (OR STFs)

The STFs are most usually revealed in EMSA analyses by means of their capacity to bind small DNA elements such as ISREs or GAS-like binding elements. Specific combinations of STFs are revealed with particular GAS elements, and until each and every GAS element has been used in an analysis of the events which follow cellular activation by a given cytokine, the whole spectrum of STFs elicited cannot be said to have been examined.

STFs ELICITED BY CLASS I CYTOKINE RECEPTORS

The structural distinction between class I and class II cytokine receptors was discussed in chapter 2. At present, there is no substantive evidence that the STFs generated by these different classes of receptor are composed of different polypeptides. However, in order to provide some structure to what is essentially a list of STFs, we have arbitrarily divided the material along the same lines as the chapter on cytokine receptors (chapter 2). In this chapter we discuss separately the STFs generated by the class I and class II cytokine receptors, and further subdivide these into STFs generated by homodimerizing and heterodimerizing class I receptors.

STFs ELICITED BY HOMODIMERIC CLASS I CYTOKINE RECEPTORS

STF-EPO/STF-TPO

STF-EPO is elicited in response to EPO and binds to the β-casein GAS element.[266-268] Composed of STAT5, the STF-EPO is indistinguishable from STF-PRL and STF-IL-3 detected with the same GAS element.[267] The PRL-GAS element binding STF-TPO complex produced in megakaryocyte lines is similarly composed of STAT5, and three other complexes were revealed on an SIE GAS element.[269,270] These were composed of STAT1 and STAT3 homo- and heterodimers. Thus, a total of four GAS binding complexes are elicited by TPO in megakaryocytes.

STF-PRL

PRL, like the interferons, triggers the activation of cytoplasmic transcription factors capable of binding to the enhancer of the IRF-1 gene (PRL-induced factor [PRLIF] and PRL-stimulated factor [PRL-SF]),[225,226,272] and a regulatory element (GAS) in the -casein gene (mammary gland factor [MGF]).[207,268] Early studies on the STFs produced following PRL stimulation of mammary cells focused upon two complexes mammary gland specific nuclear factor (MGF) and Milk Protein Binding Factor (MPBF) which resulted from EMSA using GAS elements from β-Casein[207] and β-lactoglobulin respectively.[271] These two STFs subsequently turned out to be the same, and rapid progress on MGF lead to the identification of this factor as a homodimer of a STAT-like protein.[207] In recognition of this fact, MGF was renamed STAT5. STF-PRL (PRLIF) appears to consist of two DNA-binding complexes, one of which contains the IFN-inducible STAT1 protein[225] STAT1 is also a component of the PRL-SF.[226] MGF consists of at least one STAT5 molecule; it is yet to be determined if this is a homodimer or a heterodimer. As the binding site of MGF is an imperfect palindrome (a consensus GAS element) similar to that seen in the APRF-response element,[273] the authors of this study suggest that this implies MGF may bind as a homodimer in a similar manner to the STAT3 homodimer binding the APRF-RE.[207]

STFs ELICITED BY HETERODIMERIC CLASS I CYTOKINE RECEPTORS

The involvement of several receptor chains in the binding of a particular cytokine provides an extra degree of possible complexity with respect to the generation of STFs. Thus, the availability of several receptor docking sites for signal transducers such as the STATs provides interesting combinatorial possibilities for signal outputs. Suffice it to say that the STFs detected emerging from heterodimeric receptors are only those which have been characterized using a particular GAS element as a target. However, this cannot be taken as having exhausted the possible combinations until all possible GAS elements have been explored.

STF-IL-2/IL-7/IL-15

The composition of the receptor complexes which bind to IL-2, IL-7 and IL-15 has been described in chapter. The STFs induced by IL-7 and IL-15 have been studied in peripheral blood lymphocytes (PBLs). Following either IL-7 or IL-15 simulation of PBLs, EMSA analysis using the GAS element located in the IL-2Rβ gene promoter showed an indistinguishable profile from that elicited by IL-2. Antibody band-shifted EMSAs showed that STAT5 was a component of both STF-IL-7 and STF-IL-15 in fresh PBL, while it appears to be found in complex with STAT3 in IL-2 stimulated PBLs. The similarity of these two STFs with those elicited by IL-2 can be explained by the similar composition of the receptors for these cytokines (see chapter 2).

An STF containing STAT5 is also activated in response to IL-2.[274] Induction of three IL-2RE (the GAS element of the FcγRI gene) binding STFs results from IL-2 stimulation of a transformed natural killer T cell line, YT. One of these STFs contains STAT5, while the remainder consist of STAT1 and STAT3.[274] Thus, the STFs induced by each of the members of this family of cytokines are indistinguishable. While this might seem surprising, it is worth noting that use of other GAS elements might reveal cytokine-specific STFs. As yet, these experiments have not been reported in the literature.

STF-IL-3/IL-5/GM-CSF

The STF specificity generated by different GAS elements is clearly demonstrated by the selectivity of the IL-3-induced STAT-containing complexes. IL-3 induces a number of STFs that recognize some GAS-containing elements, but not others.[228] For example, the IL-3-induced STAT5-containing STF complex that binds to the GAS element of the β-casein gene, is competed (and therefore, recognized) by the equivalent sequence present in the IRF-1, FcγRI, FcεRIIIb and APRF gene promoters, but not by the Iε, Ly6E, GBP or SIE, GAS-like elements.[228] Thus, these studies have identified the presence of an IL-3-responsive GAS-like element that must contain significant sequence differences. This IL-3-induced, β-casein GAS element-binding STF, demonstrably contains at least

STAT5A and STAT5B.[228] Similarly, IL-3 induces a STAT5A and 5B-containing factor that recognizes the GAS elements within the β-casein and IRF-1 genes, but not those present in the Ig germline ε gene promoter or Ly-6E gene promoter.[228]

STF-IL-4

STAT6 was initially identified as a component of an IL-4-induced STF.[208] The IL-4-induced DNA-binding factor, termed STF-IL-4[230] or IL-4 STAT[208] has been purified and molecularly cloned,[208] and displays a high degree of similarity with the prolactin-induced DNA-binding factor MGF, or STAT5.[207] Characterization of the STAT6-containing complexes demonstrated that the inducible factor recognizes the IL-4RE (GAS element) in the FcγRI[208] and Ig germline ε.[230] STAT6, clearly homodimerizes in response to IL-4, as demonstrated by chemical crosslinking studies.[208]

STF-IL-9

The transcription factor complexes which emerge from stimulation of the IL-9 receptor are not well understood. However, in one study undertaken in MO7E cells (a human megakaryocyte cell line), IL-9 elicits the formation of single GAS element binding STF.[275] The tyrosine phosphorylation of two STAT-related proteins, p91 and p88, suggest that this STF is a heterodimeric STAT complex. The identity of the p91 appears to be STAT1; the identity of the p88 is likely to be STAT5.

STF-IL-12

IL-12 induces two GAS binding STFs on the M67 SIE and one on FcγRI.[252] The FcγRI STF-IL-12 is a unique STAT complex consisting of STAT3 and STAT4,[252] whereas the two M67 SIE-binding STFs contained STAT1 homodimers and STAT3/STAT4 heterodimers, respectively.[252] IL-12 is the only known cytokine capable of inducing the phosphorylation of STAT4 on tyrosine,[252,276] and thus the generation of a STAT3/STAT4 heterodimer is equally unusual.

STFs Induced by the IL-6/LIF/OSM/CNTF Receptors

STF-IL-6/APRF

The stimulation of cells by IL-6 induces both the so-called SIF complexes that recognize SIE sites, and acute phase response factor (APRF), that recognizes an IL-6 response element (IL-6RE) present in the promoter of the human α_2-macroglobulin gene.[206,254,273] IL-6REs and SIE sites are highly related, and in fact both conform to the GAS consensus sequence (see Table 5). It appears that the SIF complexes contain STAT1 and STAT3 homo- and heterodimers, but APRF contains only STAT3.[206,254,281] Thus the same cytokine is capable of inducing related (perhaps identical) DNA-binding factors that selectively recognize apparently homologous response elements. Additional modifications are needed in order to rule out that these factors are in fact different in some other way. However, failing a clear-cut explanation such as phosphorylation, it remains that the distinction may be based strictly upon the subtle sequence differences between these two promoter elements.

STAT3/APRF is also involved in EGF, PDGF, and CSF-1 signaling.[197,256,277-279] The various SIF complexes are inducible by a range of cytokines[279,280] and the relative appearance, or abundance, of each of these complexes appears to depend on the level of cytokine in question. For example, at low levels of IL-6, the predominant complex induced is SIF A, and at higher concentrations of IL-6, SIF B and SIF C complexes were formed, while SIF C is also inducible by IFNγ.[256] These observations are indicators that a further level of specificity and control is built into the activation of particular genes.

In addition to the STAT-only complexes that STAT3 is involved in, IL-6 induces the cooperative binding of STAT3 with the basic-leucine zipper containing transcription factor c-Jun.[273] This composite factor is the first interaction of this kind to be documented, insofar as the STATs have hitherto been seen to associate with other nuclear transcription factors. Upon stimulation with IL-6, a STAT3 splice variant, STAT3β, associates with c-Jun to bind a composite site in the promoter of the human α_2-macroglobulin gene.[273] This composite site is comprised of an IL-6RE as described above, adjacent to a putative AP-1 element (c-Jun binding site), and, as previously discussed, recognizes the heterodimeric factor.

STFs Elicited by Class II Cytokine Receptors

STF-IFNα/β/ISGF-3

STAT1 was initially identified as a component of the IFNα-induced STF, ISGF3,[202,203,211,282] which is comprised of STAT1, STAT2 and p48.[220] ISGF3 specifically recognizes interferon-stimulated response elements (ISREs), and was first shown to require the minimal ISRE from the interferon-stimulated gene 15 (ISG15)[261] for gene transcription.

While the predominant factor induced by IFNα is ISGF3, it appears that a minority complex of STAT1 homodimers are also stimulated by IFNα, termed α-activated factor (AAF).[283] AAF is capable of binding a GAS element[283] and is indistinguishable from the IFNγ-induced GAF.[238]

The only documented utilization of STAT2 to date is its role in ISGF3, the IFNα-activated transcription complex.[203,209] As described briefly above, ISGF3 is a complex of STAT2 together with STAT1 and p48, and binds to the promoters of IFN-inducible genes containing an ISRE.[265] The stability of this complex is dependent on the presence of STAT2, and indeed, genetic evidence shows that in the absence of STAT2 (mutant fibroblasts, U6A[284]; see chapter 6), IFNα signal transduction fails to occur.[284] Studies have shown that the interaction of p48 with the STAT1/STAT2 heterodimer occurs at the point of contact with DNA, and that a complete ISGF3 complex does not exist in the absence of DNA.[265]

STF-IFNγ/GAF

STAT1 was later determined to be the sole component of the IFNγ-activated factor (GAF)[217] that binds to the gamma-activated sequence (GAS), in IFNγ-stimulated genes.[216] The definitive genetic experiments utilizing the U3A cells defective in STAT1 expression (see chapter 6), demonstrated the critical role of STAT1 in IFNα and IFNγ signaling.[285]

STF-IL-10

The IL-10 receptor is a member of the class II cytokine receptor family. Stimulation of mouse peritoneal macrophages with IL-10 lead to the rapid induction of two GAS element binding activities (specifically using the FcγRI element) with similarities to the GAS binding STFs produced in the same cells by IL-4. Nonetheless, the two sets of STFs are distinct with respect to their composition since they behave differently in EMSAs with different GAS elements, and the ability of antibodies against STAT1 (and its relatives) to disrupt the STFs generated by each cytokine was markedly different. It was deduced that the STF-IL-10 contained either STAT1, or most probably, a STAT1-like protein, although there was no speculation about what the identity of any of the other components of these complexes might be.

STFs ELICITED BY RECEPTOR TYROSINE KINASES

There is a class of inducible GAS binding factors stimulated, at least, by cellular growth factors utilizing receptor tyrosine kinases,[258,277,286] that also contained STAT1.[256,280,287] This growth factor-inducible factor, known as *sis*-inducible factor (SIF) was first identified as a transcriptional activator of the c-Fos gene through a specific *sis*-inducible element (SIE), induced by conditioned media from cells transformed with the oncogene *v-sis*.[286] It has since been shown that epidermal growth factor (EGF) induces this factor also.[288] Attempts to reveal the components of EGF-induced SIF complexes (3 in total, SIF-A, -B, and -C) demonstrated the presence of proteins similar in size to STAT1, the involvement of STAT1 in SIF-B, and SIF-C was confirmed with antibodies directed against STAT1.[256] Furthermore, the SIF-C binding activity could compete with a GAS oligonucleotide, suggesting a similarity with GAF, the STAT1 homodimeric factor.[256] As one can see from Table 5, there is a high degree of homology between the core sequence of the human *c-fos* SIE and the GAS consensus sequence, and indeed, the single point mutation in the m67 high affinity SIE renders it identical to the GAS consensus.

These experiments have since been backed up by the demonstration that the SIF complexes are composed of combinations of STAT1 and STAT3, such that the highest mobility complex, SIF-C comprises STAT1 homodimers, the next fastest SIF-B is a STAT1/STAT3 heterodimer, and finally, the slowest migrating complex is a STAT3 homodimer.[205]

DISCUSSION

Cytokines elicit both specific and pleiotrophic responses from target cells. The challenge of the last ten or so years has been to explain both of these types of response. So far, two levels of regulation of intracellular signaling have been touched upon in this book; namely, the notion that cytokine receptors recruit specific signal transduction molecules to the cell membrane, and, in this chapter, the idea that there are nuclear DNA binding complexes specifically generated by particular cytokines, which define the type of signal that ultimately arises from a given cytokine receptor. The first tier of signal specificity, that which occurs at the cell membrane, will be dealt with again in the following chapter. The latter tier of selectivity, that which defines the specificity of response at the level of the gene, remains rather enigmatic, although there are some clues which point to possible mechanisms by which both pleiotropy and specificity might be brought about. The first of these is the notion of the "half-site" of an SBE. Thus, an SBE might be composed of two complimentary half-sites, generating a palidromic SBE, or might be composed of a STAT-binding half-site combined with a half-site for a different DNA binding protein, as in perhaps the most extreme case, the ISRE. Since cytokine stimulation of a given cell results in the generation of one or a number of STF complexes, it is not difficult to account for both the pleiotropic and specific effects of cytokines at the level of gene transcription. In the case of specific gene activation, a cell type or cytokine specific STF might recognize a combination of half-sites which is a poor ligand for other STFs. Whereas, a gene which is induced by many cytokines might possess an optimal site for many STAT dimer combinations allowing its induction by a number of different STFs. The resolution of the question of specificity versus pleiotropy must wait for the analysis of the promoters of a broader range of cytokine inducible genes. Experiments using the naturally occurring SBEs in such genes should help prove or disprove this notion.

CHAPTER 6

Somatic Cell Genetic Dissection of Interferon Signal Transduction Pathways

Over the last half decade, the interferon α/β and interferon γ signaling systems have been the topic of a most elegant series of experiments. While classical biochemistry has revealed much about each of the protagonists in these pathways, their involvement and the order in which they become involved has only recently come to light from the investigation of a series of cellular mutants in the IFNα/β and IFNγ signaling pathways (reviewed in ref. 221). In this chapter we focus upon the use of somatic cell genetics as a tool with which to uncover the molecules involved in connecting the cell surface receptors for interferon α/β and interferon γ with the genes induced by these cytokines. The signaling framework which emerged from these studies has proved to be a general feature of all signaling downstream of cytokine receptors, and provided the mechanistic underpinning for the JAK-STAT pathway. The first piece in the puzzle was the generation of cell lines defective in IFNα/β signaling.

ESTABLISHMENT OF CELL LINES DEFECTIVE IN IFNα/β SIGNALING

In the mid 1980s, two groups of workers embarked on a somatic cell genetic dissection of IFN signaling, each using a different selectable element. In the most productive series of experiments, the establishment of a number of human fibroblast cell lines defective in IFNα/β[289-291] signaling was achieved by means of a selection protocol based around the use of a selectable marker (guanosine phosphoribosyltransferase [gpt] placed under the control of an IFNα/β-dependent promoter [from the 6-16 gene]).[292] The details of this selection protocol are outlined in Figure 17 (A,B,C). Briefly, IFNα/β treatment of HT-1080 human fibroblast cells transfected with this construct (a cell line now called 2fTGH) resulted in the induction of the 6-16 promoter, and in consequence the expression of gpt. These cells lack hypoxanthine phosphoribosyltransferase (HPRT), and therefore in the presence of IFNα/β cannot grow in hypoxanthine-aminopterin-thymidine (HAT), while being resistant to the effects of 6-thioguanine (6-TG). IFNα/β-induced expression of gpt resulted in the recovery of the ability to grow in HAT, and the acquisition of 6-TG sensitivity. Thus, in the $HPRT^-$ cell line 2fTGH, both positive and negative selections could be made for the viability of the IFN pathway, simply by assaying for growth in HAT or in the presence of 6-TG, with or without treatment with IFNα/β. By serially treating the parental 2fTGH cell line to 70% lethality with the frameshift mutagen ICR-191, a series of cell lines which were defective in IFNα/β signaling were generated. In these mutant cell lines, the ability to grow in HAT medium in the absence of IFNα/β was now abolished, although they remained insensitive to the effects of 6-TG. Complementation of these cell lines with cDNAs encoding the genes which had been rendered inactive by the mutagen treatment would rescue this phenotype, re-establishing IFNα/β-dependent growth in HAT and 6-TG sensitivity. The mutant cell lines so obtained were classified into six complementation groups (U1 to U6; reviewed in ref. 221) based upon the ability of members of each group to form heterokaryons which complement the defects in each other's IFNα/β signaling system.

ESTABLISHMENT OF CELL LINES DEFECTIVE IN IFNγ SIGNALING

For IFNγ, two independent studies were undertaken. In the first instance, an IFNγ-dependent promoter (from the 9-27 gene[293]) was used to drive the cell surface protein CD2, which in turn permitted selection of the mutant cells by fluorescence activated cell sorting (FACS).[294,295] Again, a round of serial mutagenizations of the parental fibroblast line was undertaken.

In the alternate series, a second group of investigators used a procedure similar to that used for the IFNα/β pathway, followed for IFNγ alone this time, with an MHC class II promoter from the HLA-DRα gene driving a herpes simplex virus thymidine kinase (HSV-TK) gene for metabolic selection in HeLa cells.[296] This system also provided a secondary selection by FACS for those cells that failed to induce MHC class II on the cell surface. These latter experiments defined two classes of IFNγ signaling mutant (A, B). Interestingly, class B mutants were also defective in IFNα/β dependent induction of several mRNA species, demonstrating functional overlap between the two pathways.

Thereafter, in both studies, the task of defining the genetic defect in each of these complementation groups was undertaken, and the data generated has been pivotal in the ordering of each of the components of the JAK-STAT signal transduction pathway. The cellular mutants and the genes which complement them are shown in Table 6.

Table 6. Complementation of IFN signal transduction mutants

Group	Defect	Complementation	Reference
U1	IFNα	TYK2	199
U2	IFNα/β, IFNγ	ISGF3γ (p48)	Described in 221,285
U3	IFNα/β, IFNγ	STAT1	285
U4	IFNα/β, IFNγ	JAK1	300
U5	IFNα/β	IFNaR2	156
U6	IFNα/β	STAT2	284
γ1	IFNγ	JAK2	295
γ2	IFNγ	???	

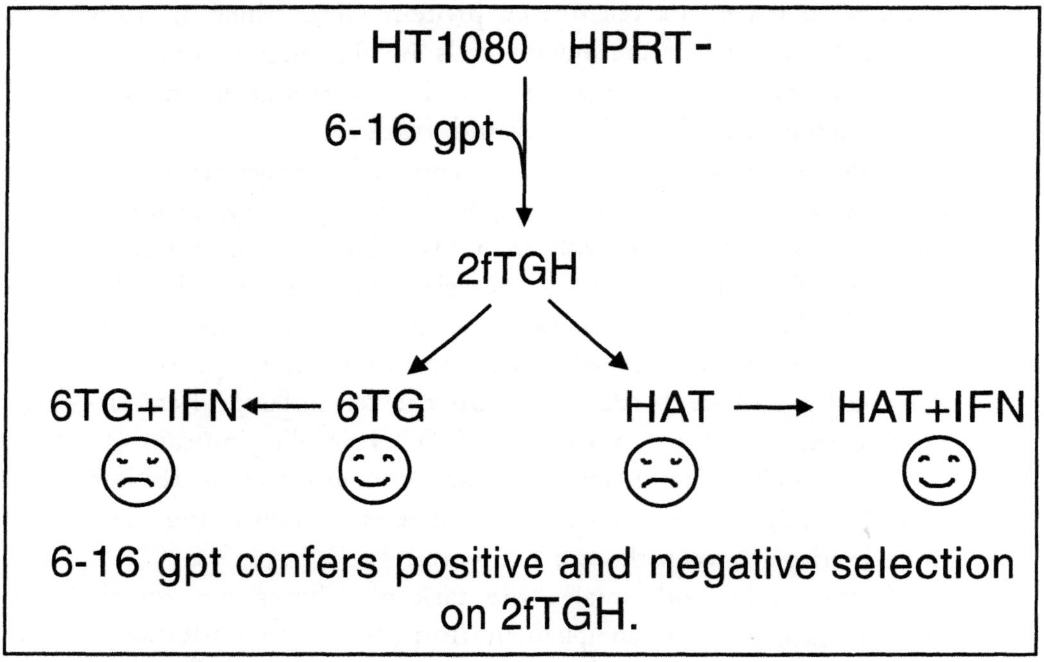
HT1080 HPRT-
6-16 gpt
2fTGH
6TG+IFN
6TG
HAT
HAT+IFN
6-16 gpt confers positive and negative selection on 2fTGH.

A

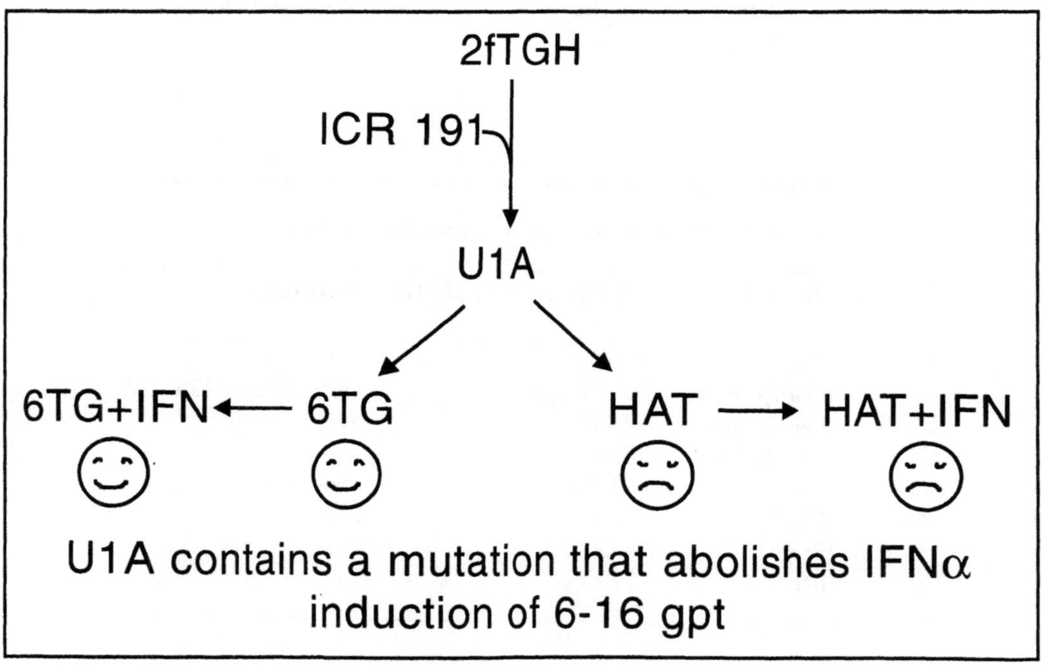
2fTGH
ICR 191
U1A
6TG+IFN
6TG
HAT
HAT+IFN
U1A contains a mutation that abolishes IFNα induction of 6-16 gpt

B

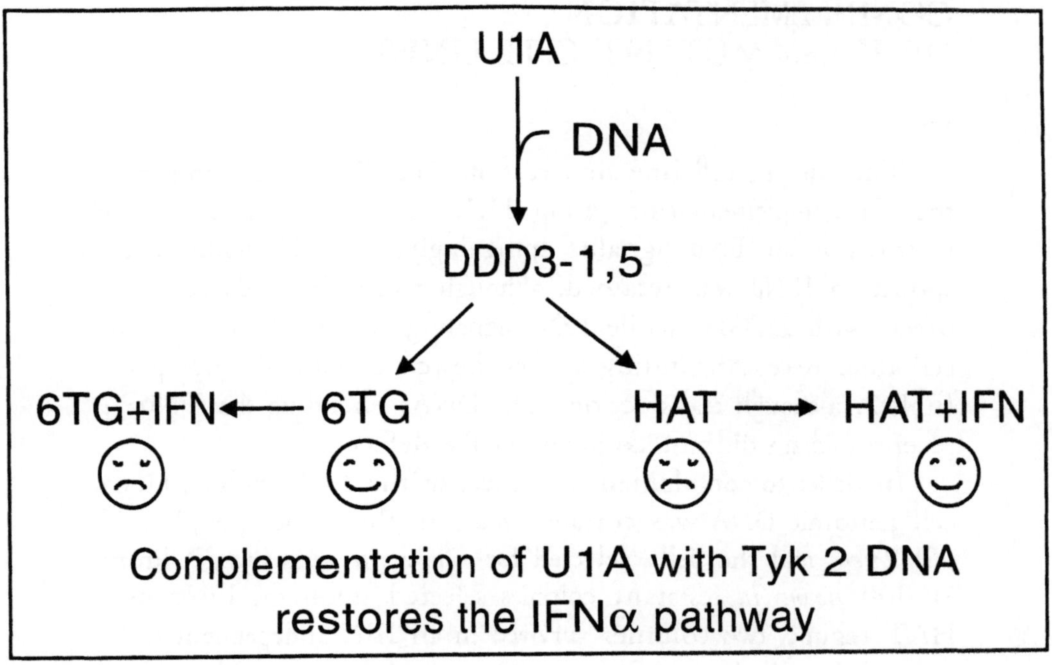

C

Fig. 17. Schema depicting the generation of the U1-U6 mutant fibroblast cell lines, and identification of the mutant components created therein.

Panel A. The HPRT™, human fibrosarcoma cell line, HT1080 was stably transfected with the guanosine phosphotransferase (gpt) gene under the control of the IFNα-inducible promoter from the 6-16 gene. The resultant cell line, 2fTGH, was capable of expressing gpt upon stimulation with IFNα, and was sensitive to the presence of a mixture of hypoxanthine, aminopterin and thymidine (HAT). In the presence of IFNα, and hence the induction of the expression of gpt, these cells were now rendered insensitive to HAT [HAT + IFN], and sensitive to 6-thioguanine (6TG) [6TG + IFN].

Panel B. 2fTGH cells were then treated with the point mutagen ICR 191, in order to generate cell lines no longer capable of transducing an IFNα signal. This mutagenization resulted in cells now insensitive to 6TG in the presence of IFNα [6TG + IFN] and sensitive to HAT in the presence of 6TG [HAT + 6TG]. Thus the expression of gpt was no longer inducible by IFNα, demonstrating the presence of mutation(s) within one or more components of the IFN signaling cascade.

Panel C. In order to identify these mutated components, the first mutant complementation group, U1A (defective in signaling both IFNα/β and IFNγ) was subjected to multiple rounds of human genomic DNA transfections, until the cells were restored to wild type with respect to IFNα signaling. These cells are now capable of transducing an IFNα signal as defined by the HAT and 6TG selection criteria for the 2fTGH cells. (Figure adapted courtesy of A. Oates).

COMPLEMENTATION OF IFNα/β MUTANT CELL LINES

U1

The mutant cell line derived from 2fTGH and a member of the U1 complementation group, 11.1, was defective in its ability to transmit an IFNα signal.[297] Intriguingly, the cell's ability to respond to IFNβ was retained, although somewhat reduced compared with 2fTGH, while IFNγ signaling was unaffected in this cell line. Receptor binding studies showed a reduced binding for IFNα2, although transfection with cDNAs encoding the IFNαR1 receptor chain did not complement the defect in 11,1.

In order to complement the defect in the 11,1 cell line, HeLa cell genomic DNA was cotransfected into them with the plasmid pSV2*neo*, and the cells selected for *neomycin* resistance.[298] From 10,000 *neomycin* resistant colonies selected upon the IFNα plus HAT regime, two colonies survived in an IFN independent fashion, while a single transfectant (code-named D3-1) was dependent upon IFN for survival in the presence of HAT, and was therefore a strong candidate for complementation of the IFN pathway defect. Again, DNA from the primary transfectant D3-1 was prepared, and a second round of transfection and selection in *neo* was undertaken, this time resulting in seven colonies, of which only one (DD3-1) survived in IFNα plus HAT. A third round of transfection was undertaken using DNA from DD3-1, this time fourteen *neo*-resistant colonies were obtained, and again one line (DDD3-1,5) possessed the required IFNα sensitive phenotype. At this point, it was felt that in DDD3-1,5 the *neo* marker was segregating with the gene which was complementing the defect in IFNα signaling. Finally, a cosmid library was generated from DDD3-1,5, and selection for *neo*-resistance in the bacterial transformants demonstrated that the same fragment of human DNA had been captured by the pSV2*neo* plasmid. Surprisingly, this fragment of DNA contained a portion of the gene encoding the JAK family member Tyk2,[180] and suggested that this gene encoded the protein defective or missing in the pathway downstream of the IFNα receptor. A final round of transfection of 11,1 cells with cosmids encoding the entire Tyk2 gene were able to complement the IFN signaling defect in 11,1, and firmly placed this PTK in

IFN signaling. Finally, the complementation of 11,1 resulted in the restoration of high affinity binding of IFNα on these cells, suggesting that this was a genuine consequence of the lack of Tyk2 in these cells, rather than the lack of one of the ligand binding components of the receptor.

Thus, after five years of Herculean effort, rescue of the first complementation group by the JAK family member Tyk2[299] opened the flood gates, and in record time JAK1,[300] JAK2,[295] STAT1[285] and p48[290] were implicated in this pathway by virtue of their ability to rescue cell lines from each of the particular complementation groups generated in the screen, or by biochemical analysis of defective cells.

U2

The U2 complementation group was generated in the same series as the U1 cell line. It was subsequently characterized in terms of its response to IFNα, IFNβ and IFNγ, both at the level of IFN-responsive gene regulation and with respect to its antiviral response to IFNγ.[290] The U2 cell line was unable to activate any of the nine IFNα/β responsive genes tested, while genes inducible by IFNγ appeared largely unaffected. One exception was the 9-27 gene which lost its inducibility by both IFNα and IFNγ. Interestingly, the U2 cell line also had a greatly diminished IFNγ antiviral response. EMSA analysis demonstrated that one component of the IFNα/β inducible transcription factor ISGF3 was missing. Known at the time as ISGF3γ, p48 appeared to be absent from this mutant line, and its absence resulted in the inability to form ISGF3, the STAT1/STAT2/p48 containing complex which mediates the IFNα/β transcriptional response. Perhaps the most fascinating property of this mutant cell line is its reduced IFNγ antiviral response, providing genetic evidence for the involvement of p48 in the IFNγ response. This effect was not readily seen in the mere biochemical analysis of transcriptional responses in this, or any other, cell line. At present, the complementation of the U2 cell line with a cDNA encoding the p48 gene has not been published. However, there are two citations of "unpublished data" which purport to have successfully achieved this,[221,285] and these seem to be proof enough that it is indeed the case.

U3

Members of the U3 complementation group were completely defective in transcriptional response to both IFNα/β and IFNγ.[285] Detailed analysis of the U3A cell line showed that two components of the transcription factor ISGF3 were missing. These proteins, at the time known as p91 and p84 but now known as STAT1α and STAT1β, respectively,[202,211] are differentially spliced products of the same gene, and are identical except for the absence in STAT1β of the carboxy-terminal 38 amino acids present in STAT1α.[202] Interestingly, whereas a cDNA encoding STAT1α is able to complement the defect in both IFNα/β and IFNγ signaling, STAT1β cDNA is only able to complement the IFNα/β signaling defect. Since both STAT1α and STAT1β are able to participate in the formation of transcription complexes such as ISGF3, the inability of STAT1β to complement the IFNγ pathway suggests that there are domains crucial to IFNγ signaling in the carboxy-terminal region of STAT1 which are required for some signaling mechanisms, but not others. Intriguingly, buried in the 38 amino acids absent from STAT1β is a serine phosphorylation site required for IFNγ signaling.[241] It seems likely that this may be the basis of the differential potency of STAT1α and STAT1β in the complementation of the U3A cell line.

U4

Members of the U4 complementation group are defective in both IFNα/β and IFNγ signaling.[300] Although diminished binding of IFNα/β was reported for the U1A cell line, it is not clear whether this is also the case with the U4A cell line. Western blot analysis of detergent extracts of the U4A cell line showed that JAK1 was absent, while Northern blot analysis revealed the presence of a truncated JAK1 transcript. Further, it was demonstrated that the tyrosine phosphorylation of JAK1 and Tyk2, and of JAK1 and JAK2, characteristic of the intracellular responses to IFNα/β[203,211,282,284,285,301-308] and IFNγ,[217,249,250,285,309-314] respectively, were both missing in the U4A cell line. Similarly, neither STAT1 nor STAT2 became tyrosine phosphorylated in response to IFNγ or IFNα/β. All of these defects were corrected by reintroduction of a functional JAK1 cDNA.[285]

U5

The U5 complementation group is the most recently characterized of the IFN signaling pathway mutant cell lines.[156] The cell line is unable to respond to IFNα/β, however, it is fully responsive to both IFNγ and IL-6. The cell line is unable to bind IFNα/β and was shown to be defective in the cell surface expression of one of the components of the IFNα/β receptor (known as ifnar2[315]). This receptor chain is found as both a long form (ifnar2-2)[156] and a short form (ifnar2-1),[151] each of which is a product of differential splicing from the same gene.[156] In contrast to the ifnar-2-1 form of this IFNα/β receptor chain, the long form possesses an intracellular domain of 251 amino acids which bears a Box 1/Box 2 region of the type held in common by many cytokine receptors (also see chapter 2). Significantly, only a cDNA encoding the ifnar2-2 was able to complement the U5A cell line, suggesting that the ifnar2-1 is lacking one or more of the key intracellular elements required to recruit the appropriate intracellular signals in response to IFNα/β.

U6

The U6 complementation group is defective in IFNα/β signaling, while retaining a fully functional IFNγ signaling pathway.[284] Western blot analysis revealed that the cell line U6A lacked detectable levels of STAT2, and was unable to assemble an ISGF3 complex in response to IFNα/β. Importantly, STAT1 tyrosine phosphorylation in response to IFNα/β was demonstrated to be dependent in large measure upon the presence of STAT2, hinting at the mechanism by which STAT1 is recruited to the ISGF3 complex (see chapter 7). Successful complementation of this cell line with a full-length cDNA encoding STAT2 restored all of the defects in the U6A cell line, including the recruitment of STAT1 phosphorylation.

COMPLEMENTATION OF IFNγ MUTANT CELL LINES

γ1

The second series of IFN signaling pathway mutants (including γ1 and γ2) were generated (as described above) by FACS selection of mutagenized fibroblasts expressing the cell surface protein CD2 expressed under an IFNγ-dependent promoter (from the 9-27 gene[293]). Members of the γ1 complementation group lacked an IFNγ response, although the IFNα/β pathway appeared to be intact.[295] Analysis of the expression levels of each of the likely suspects (STAT1, JAK1, JAK2 Tyk2, p48 etc.) revealed no obvious defects in their mRNA expression. However, the tyrosine phosphorylation of JAK2 and JAK1, which is characteristic of the response of these cells to IFNγ, was absent in these cells. Since the IFNα/β dependent tyrosine phosphorylation of JAK1 and Tyk2 appeared to be normal, suspicion fell upon JAK2 as the most likely candidate for the γ1A defect. Complementation of γ1A by a cDNA encoding JAK2 demonstrated that this was indeed the case. The precise nature of the defect in JAK2 is at present unknown, however, it seems likely that a point mutation or small deletion in the coding sequence is the cause of the problem.

γ2

Members of the γ2 complementation group are defective in IFNγ signaling.[221] The cells do, however, retain a functional pathway for IFNα/β signaling. The molecular basis of the defect is at present not understood, and there is no published record of the complementation of a γ2 cell line.

ORDERING OF THE JAK-STAT PATHWAY USING IFN SIGNALING MUTANTS

The availability of the mutant cell lines described in this chapter represented a major step forward in our understanding of IFN signaling. Their successful complementation by cDNAs encoding many of the components of the JAK-STAT pathway was much more than a mere confirmation of the biochemical studies which were being undertaken on these molecules at that time. Rather, the experiments described above were genetic proof of what could only be inferred from the biochemistry, and in this way established a foundation upon which the whole JAK-STAT pathway could be built. Moreover, these cell lines could now be used to probe the finer points of the JAK-STAT pathway, such as the interdependence of individual components of the pathway upon each other.

Thus, by studying other components such as JAK2, Tyk2, STAT1 and STAT2 pathway in a background devoid of functional JAK1 (U4A cells for example), the reciprocal interdependence of Tyk2 phosphorylation upon JAK1 in IFNα/β signaling became apparent. Similarly, a dependence of STAT1 phosphorylation upon JAK1 was observed in these cells, suggesting that a functional IFNα/β receptor complex needed both JAK1 and Tyk2 in order to be capable of transmitting a signal via STAT1 and STAT2. More recently, in the IFNα/β pathway, the requirement of STAT2 phosphorylation, prior to (and indeed obligatorily for) STAT1 phosphorylation suggests that a particular order of events must occur for the generation of the correct IFNα/β specific signal. This ordered assembly of the STAT complex on the receptor perhaps explains how the selection of the precise components of the STAT complex is determined in response to a particular physiological trigger. As already noted, Tyk2 also appeared to be required for the generation of a stable IFNα/β binding site upon the surface of the U1A mutant cell lines, suggesting that it is required for the formation of a pre-existing multimeric receptor complex. Taken together with the biochemical analyses undertaken upon the JAKs and STATs (and discussed in the sections above), the somatic cell genetic approach has both confirmed and extended the predicted sequence of events in the JAK-STAT pathway.

CHAPTER 7

A Guided Tour Through the JAK-STAT Pathway

INTRODUCTION

In previous chapters we discussed the properties of each of the classes of proteins involved in the JAK-STAT pathway. Armed with this information, it is now appropriate to move on to a discussion of the pathway itself. In doing so, the pathway has been stripped of any of the loose ends which inevitably adorn a cellular mechanism of such recent discovery. We are aware that, in discussing only the general principles of the JAK-STAT pathway, we are presenting an incomplete picture. However, this, combined with the material in chapter 9, presents an idea of how the JAK-STAT pathway blends into the overall view of signal transduction in mammalian cells.

The general principles of the JAK-STAT pathway are presented for IFNγ signaling in Figures 18-24. The same basic mechanism underlies the signal transduction processes initiated by virtually all other cytokine receptors. Thus, the JAK-STAT pathway is demonstrably activated in response to erythropoeitin,[266,268,316-319] G-CSF,[320-323] growth hormone,[268,304,311,324-326] prolactin,[311,327-330] thrombopoietin,[269,270,331-333] ciliary neurotropic factor,[123,222,312,334-339] IL-6,[273,310,321,334,336,337,340-347] IL-11,[348,349] LIF,[123,334,336,350] oncostatin M,[350] GM-CSF,[228,263] IL-3,[228,229,263,321,333,351] IL-5,[228,229,352] IL-2,[145-147,266,274,302,318,353-358] IL-4,[146,208,231,263,355-365] IL-7,[146,355,358,366-368] IL-9,[275,361,369] IL-12,[252,253,370] IL-13,[146,358] IL-15,[146,365,371,372] IFNα/β,[203,211,282,284,285,301-308] IFNγ[217,249,250,285,309-314] and IL-10.[373] However, there remain discrepancies in detail for each specific cytokine induced pathway which are unresolved. These will be addressed in subsequent sections. The JAK-STAT pathway can be conveniently divided into eight distinct events, namely:

(1) Ligand induced cytokine receptor dimerization;
(2) Cytokine receptor tyrosine phosphorylation;
(3) Docking of STATs onto cytokine receptors;
(4) Tyrosine phosphorylation of STATs;
(5) Formation of functional STAT dimers;
(6) Serine Phosphorylation of STATs;
(7) Nuclear translocation of STATs and activation of genes by STAT containing DNA binding complexes; and
(8) Tyrosine phosphatase mediated resetting of the pathway.

Each of these steps is outlined in Figures 18-24, and each will be discussed in depth.

LIGAND INDUCED CYTOKINE RECEPTOR DIMERIZATION

The activation of the JAK-STAT pathway has its origins with the binding of an extracellular ligand (or cytokine) to its cognate receptor. Ligand binding leads to the oligomerization of two (or more) membrane-spanning receptor subunits, resulting in the formation of an 'activated' transmembrane receptor complex. The best documented receptor/ligand complex is the growth hormone/growth hormone receptor (GH/GH-R) complex, the structure of which has been deduced by X-ray crystallography.[102] In this case, the growth hormone receptor (GH-R) is induced to homodimerize by dint of binding of a single growth hormone (GH) molecule to distinct regions on each of the receptor chains.[41] The dimerization of the external domains of the GH receptors results in the enforced proximity of the intracellular domains of each receptor chain. This effectively converts the extracellular signal into an activated receptor complex capable of transmitting the signal into the intracellular milieu. A series of intracellular events follows the generation of the activated receptor complex, the first of which is the activation (or, more accurately, the recruitment and subsequent activation) of receptor-'associated' PTK activity.

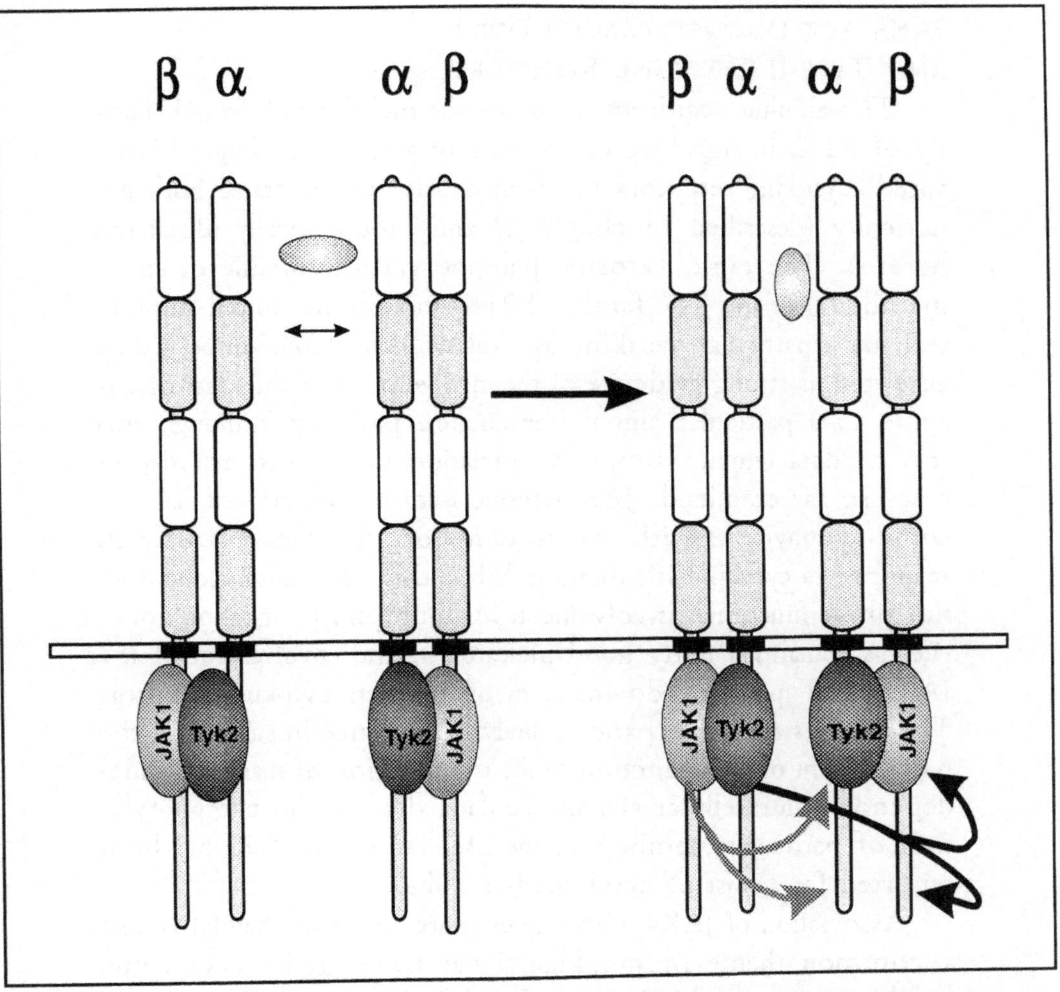

Fig. 18. A receptor chain is substrate for the PTK activity of JAK. The IFNα/β receptor complex is triggered to activate JAK PTK activity by means of receptor interaction. The intracellular domain of the receptor is a substrate for this PTK activity.

JAKs Act Downstream of Type I and Type II Cytokine Receptors

The absolute requirement for specific members of the JAK family of PTKs in signaling downstream of a number of ligand 'activated' cytokine receptors has been clearly demonstrated both genetically (described in chapter 6) and biochemically (discussed herein). The rapid tyrosine phosphorylation of one or more member(s) of the JAK family of PTKs in response to cell stimulation by a particular cytokine (see Table 7), has come to be widely accepted as strong evidence of the involvement of this class of enzyme in a particular signal transduction pathway. Although this sort of data remains simply a correlation in the vast majority of cases so far examined, JAK tyrosine *auto-phosphorylation* has become synonymous with the *activation* of JAK kinase activity in response to cytokine stimulation. While this correlation alone does not prove functional involvement, JAK autophosphorylation nonetheless remains a fairly good indicator of the involvement of the JAK-STAT pathway downstream of a given cytokine receptor. There exists an ever expanding body of evidence in support of the requirement of JAK function in the transduction of many cytokine-dependent intracellular signals. To date, the tyrosine phosphorylation of particular members of the JAK family of PTKs has been reported for a host of cytokines (see Table 7).

Association of JAKs with one or more receptor chain(s) is also a common theme in cytokine signal transduction. For some cytokine receptors this has manifested itself as a constitutive association (as appears to be the case for JAK1 and the "common" chain of the IFNα/β-R,[151,307] IFNγ-R,[313,374] gp130[334] and G-CSF-R[320] inter alia, and for JAK2 and EPO-R,[316] G-CSF-R[320] and PRL-R[328] inter alia. In contrast, however, ligand-inducible recruitment of JAK2 appears to be the case for the GH-R.[324]

There are defined sites of association for members of the JAK family on particular cytokine receptors. For example, the proline-rich motif within Box 1 (see chapter 2) appears to be necessary for the association of JAK2 with the GH-R,[104,375] PRL-R,[329,376] G-CSF-R[320] and EPO-R.[375] In the IFNγ system, however, JAK association sites are somewhat different. For the IFNα-receptor α-chain, the sequence (-LKPS-) located between residues 266 and

Table 7. Compilation of the JAKs and STATs activated by a range of cytokines

Cytokine	Subunit Structure	JAK: JAK1	JAK2	JAK3	Tyk2	STAT: STAT1	STAT2	STAT3	STAT4	STAT5	STAT6	References
EPO	Homodimer		■							■		266,268,316-319
G-CSF		■	■			■		■				320-323
GH			■			■		■		■		268,304,311,324-326
PRL		■	■			■				■		311,327-330
TPO			■			■		■		■		269,270,331-333
CNTF	Heterodimer of α-chain with gp130	■	■		■	■		■				123,222,312,334-339
IL-6		■	■		■	■		■				273,310,321,334,336,337,340-347
IL-11		■	■		■	■		■				348,349
LIF		■	■		■	■		■				123,334,336,350
OSM		■	■		■	■		■				350
GM-CSF	Heterodimer of α-chain with KH97/AIC2B		■							■		228,229,263,268,474
IL-3			■							■	■	228,229,263,321,333,351
IL-5			■			■				■		228,229,352
IL-2	Heteromultimer of α-chain with IL-2Rγ-chain (γC)	■		■		■		■		■		145-147,266,274,302,318,353-358
IL-4		■		■						■	■	146,208,231,263,355-365
IL-7		■		■		■		■		■		146,355,358,366-368
IL-9		■		■	■					■		275,361,369
IL-15		■		■						■		146,358
IL12	?		■		■			■	■			252,253,370
IL-13	Heterodimer										■	146,365,371,372
IFNα/β	Heteromultimer of at least α and β-chains		■		■	■	■	■				203,211,282,284,285,301-308
IFNγ		■	■			■						217,249,250,285,309-314
IL10		■			■	■		■				373

A shaded box indicates that a particular JAK or STAT is implicated in the signal transduction processes downstream of a given cytokine. The references cited contain the information from which the table was derived.

269 appears to be required for JAK1 binding,[314] whereas JAK2 requires a proline-rich element (-PPSIPLQIEEYL-) for binding to the associated β-chain[314]. Intriguingly, the association of JAK1 and JAK3 with the β-chain and the γ_c chain, respectively, of the IL-2 receptor system also requires distinct domains. JAK1 interacts with the membrane proximal region of the β-chain of the IL-2R similar to that described above for other cytokine receptors,[149] whereas JAK3 associates with a carboxy-terminal element on the γ-chain.[149]

Since the amino acid conservation found in the Box 1 and Box 2 homology domains is relatively low, consisting as it does simply of a "proline-rich" element, it suggests that the actual interactive surfaces of the JAK and the receptor may well be subtly different for each cytokine receptor and each JAK. At present, therefore, there is no coherent notion of why different JAKs associate with particular cytokine receptors, nor indeed, does one readily emerge from an analysis of the information presented in Table 7.

One contentious issue is whether or not the JAK family member(s) is constitutively associated with the receptor or whether it is recruited at the time of receptor activation. For example, in the case of the association of JAK2 with the GH-R,[324] there does not appear to be a constitutive binding of this PTK to the intracellular domain of the receptor, rather it recruits this molecule upon stimulation with GH.[324] In this respect, GH-R differs from most other homodimerizing cytokine receptors such as EPO-R,[316] G-CSF-R[320] and the PRL-R,[328] where JAK2 appears as a constitutively associated signaling chain. Nonetheless, triggering of dimerization of the GH-R by GH, causes JAK2 to become phosphorylated upon tyrosine and simultaneously (it appears) becomes activated with respect to its own intrinsic PTK activity. Similar activations are reported for other JAK family members in other cytokine receptor systems. It is worth noting from data emerging from the IFN signal transduction mutants discussed in chapter 6 that the absence of Tyk2 results in the downregulation of high affinity binding sites for IFNα,[298] suggesting that there is an important link between JAK functionality and the affinity of a cytokine receptor for its ligand. The physical association of the JAK and the cytokine receptor appears to be the most straightforward construction which can be placed upon these data, however, there are other possible explanations.

One interesting discovery has been made in the interaction of IFNγ with its multi-chain receptor complex. Here, JAK1 appears to be associated with the α-chain of the receptor,[313,374] while JAK2 must be recruited to the receptor complex for a signal to be transmitted.[374] The complex nature of the multimeric IFNγ receptor complicates the interpretation of this observation. However, since the recruitment of JAK2 might occur as a consequence of the ligand induced association of the two (or more) receptor chains involved in the binding of IFNγ, it may well be that JAK2 is preassociated with the β-chain of the IFNγ receptor, and that the act of dimerization results in the de facto association of JAK with the IFNγ β-chain.

CYTOKINE RECEPTOR TYROSINE PHOSPHORYLATION

For cell surface receptors which themselves possess PTK activity (RTKs), the initial phase of ligand-mediated receptor activation results in the elevation of the levels of receptor-associated protein tyrosine kinase activity. In turn, this elevated PTK activity brings about the phosphorylation of specific tyrosine residues within the intracellular domain of the particular receptor. While this mechanism of receptor activation has been well documented for RTKs, it also holds for the non-PTK receptors of many cytokines and this tyrosine phosphorylation plays a fundamental role downstream of cytokine receptor activation.[109] This phosphorylation is most likely to be mediated via receptor associated-cytoplasmic PTKs, and there is a growing literature implicating the JAK family of PTKs in this process. While the evidence is largely circumstantial in the majority of cases, by sheer weight of numbers (see citations in this chapter and references presented in Table 7) the case for the involvement of the JAK family of PTKs in this process is very strong. Therefore, with the caveat that the combined weight of the evidence presented herein falls marginally short of actual proof, we will now continue to build our favored model for cytokine receptor mediated signal transduction.

A Receptor Chain is Substrate for the PTK Activity of JAK

At present, our paradigm for RTK signal transduction is based upon the transphosphorylation of homodimerizing receptor intracellular domains by the intrinsic catalytic activity possessed by the receptor chains themselves. The decoration of these intracellular domains with phosphotyrosine residues initiates the association of second wave signal transducers such as SHC, GRB2 and PI-3' kinase.[14,377,378] These concepts have been elaborated upon elsewhere in this book (see chapter 1) and in other reviews (for example, see refs. 9, 378). It is now clear that a fundamentally similar model holds for the non-PTK cytokine receptors. Tyrosine phosphorylation sites have been located on intracellular domains of several cytokine receptors,[112] and for the most part these are canonical with known SH2 domain binding sites.

At present, the most thoroughly investigated system in this regard is the human IFNγ-R α-chain where a tyrosine residue located at position 440 (Y440)[379] has been shown to be phosphorylated in response to IFNγ.[246] This same site has been shown to bind STAT1 in vitro in a phosphotyrosine-dependent fashion,[246] an observation consistent with the notion that the mechanism by which STAT1 becomes associated with the IFNγ-R pathway may be through its recruitment to the receptor α-chain which had been phosphorylated by JAK1 or JAK2.

The evidence that it is the JAK kinase(s) itself that phosphorylates the receptor is at present scant, however, genetic analysis of the IFNγ system is at least consistent with the notion that without a functional JAK being present, the STAT does not become phosphorylated upon tyrosine.[295,300] Similar candidate tyrosine phosphorylation/SH2 binding motifs have been found in many other cytokine receptors, and in several cases these have been shown to be necessary for signaling[112,113,208] inter alia.

Specific cytokine stimulation-dependent tyrosine phosphorylation sites have been proposed for a number of cytokine receptors. One interesting piece of work has been that undertaken on the IFNα/β receptor, where a likely tyrosine phosphorylation/SH2 binding site has been deduced from first principles.[115] This tyrosine is located at amino acid 466 (hence its name, Y466) of the interferon-α receptor 1 subunit (IFNαR1). This site is phosphorylated in vitro by Tyk2 and binds to STAT2 but not other STATs, suggesting that this phosphotyrosine is responsible for the recruitment of this STAT to an IFNα/β-activated receptor complex.

In the case of the prolactin receptor, a single phosphotyrosine residue appears to be responsible for the activation of prolactin dependent gene induction.[113] In this study, mutation of each of the tyrosine residues in the intracellular domain of the human prolactin receptor, followed by the analysis of the ability of these mutants to induce transcription from the β-casein promoter, lead to the conclusion that all but one of these tyrosine residues (the most C-terminal tyrosine residue, located at position 580 in the long form of the receptor) were dispensable for this induction. The conclusion drawn from these experiments was that the induction of activated STAT complexes required the presence of this receptor tyrosine, presumably in a phosphorylated state. Taken together with the studies undertaken on the IFN receptor systems, there are now strong grounds for believing that the first phase of cytokine-mediated receptor activation is the generation of JAK-mediated tyrosine phosphorylated STAT binding sites on the intracellular domain of the receptor chain.

It is an attractive hypothesis that the recapitulation of the specificity of the extracellular binding of a ligand to a receptor ligand binding domain, occurs by covalent modification of the intracellular domain of the receptor chain. Moreover, the notion that a matrix of specific phosphotyrosine-bearing SH2 domain binding sites (each with its own binding specificity) are generated upon the intracellular domain of a given receptor, is one step in the important process of signal integration, which is crucial if the full range of cytokine specific responses are to emerge from an activated receptor.

Insulin Receptor Substrate-1 (IRS-1) and Insulin Receptor Substrate-2 (IRS-2) are Substrates for Activated JAKs

The tyrosine phosphorylation of IRS-1 has been shown to occur after the stimulation of cells with a number of different growth factors, prominent amongst them insulin,[380] growth hormone,[381,382]

LIF[382] and IFNγ.[382] IRS-1 is an approximately 130 kDa protein which does not appear to possess any recognizable catalytic domain (such as PTK, GTPase or PI-3' kinase activities), rather this molecule appears to function as a substrate for the PTK activities stimulated by ligand binding of these receptors. The phosphotyrosine residues generated on IRS-1 appear to be responsible for the recruitment of other SH2-bearing signal transduction molecules. Thus IRS-1 seems likely to act as a mechanism by which the SH2/phosphotyrosine mediated recruitment of second wave signaling molecules is brought about, and may broaden the scope of the signal transduction pathways activated by any given growth factor. The IRS-2 protein,[383] also known as 4PS, has recently been identified and seems to be a major substrate for the PTK activity(s) recruited by IL-4,[384] , IL-9[369] and IL-13.[384]

These molecules are of particular potential importance to the JAK-STAT pathway, since it now seems likely that all of the potential phosphotyrosine residues on the intracellular domain of the growth hormone receptor are dispensable for the recruitment of STATs.[385] The likelihood is that the recruitment of STATs to the GH response occurs through the involvement of a phosphorylated IRS-1 intermediary. More work needs to be done in other systems in order to determine whether this is a general phenomenon in cytokine signal transduction.

DOCKING OF STATs ONTO CYTOKINE RECEPTORS

As described above, the tyrosine-phosphorylated intracellular domain of the activated cytokine receptor is demonstrably the site of STAT recruitment. This recruitment mechanism is thematically consistent with all of the other signal transduction processes which are mediated by SH2/phosphotyrosine interactions (see also chapter 1). Thus, the interaction of a STAT SH2 domain with a particular receptor-borne phosphotyrosine motif is the first specificity control point of the JAK-STAT pathway.

STATs Bind to Receptor Phosphotyrosine Residues

The unveiling of the transcriptional activation potential of STAT proteins involves their recruitment to the activated receptor.

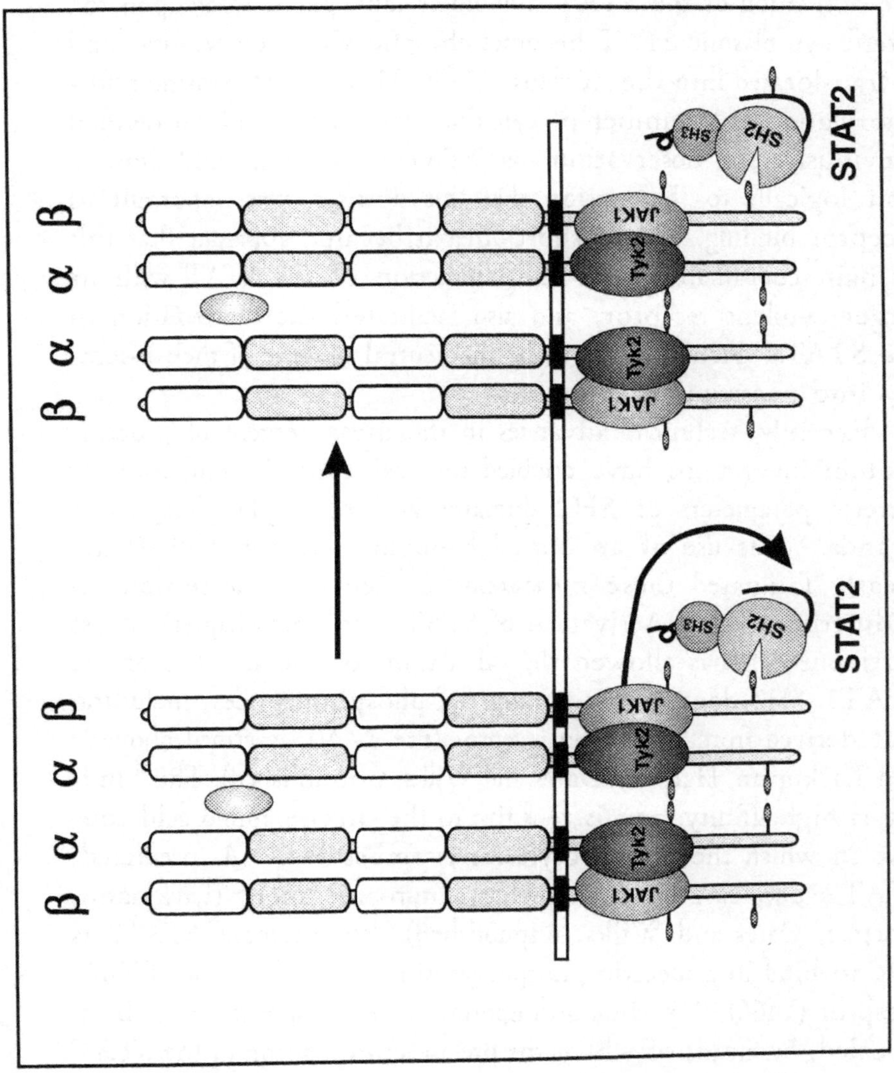

Fig. 19. Docking of STATs onto cytokine receptors and tyrosine phosphorylation of STATs. The tyrosine phosphorylated intracellular domain of the IFNα/β receptor possesses docking sites for STAT2. The proximity of the STAT protein to the activated PTK activity of JAK1 and Tyk2 results in the phosphorylation of the STAT2.

It is here that the molecular interactions which are responsible for the activation of the STAT monomers take place, whereupon the latent cytoplasmic STAT becomes phosphorylated on tyrosine, and is translocated into the nucleus.[211,217,219] The sites of tyrosine phosphorylation of a number of cytokine receptors have been defined previously. The observation that STATs possess an SH2 domain lead logically to the notion that this domain was important to receptor binding, and in short order it became apparent that this domain coordinated both the interaction of the STAT with its target cytokine receptor, and also facilitated the dimerization of the STATs which appears to be the central feature of their assembly into transcription complexes.

Recently, technical advances in the measurement of protein-protein interactions have enabled the real time determination of kinetic parameters of SH2 domains and their phosphotyrosine ligands.[13] The use of an optical biosensor (BIAcore™)[314,386] has greatly facilitated these measurements (see also the section on "Dimerization and Activation of STATs" in this chapter). Most pertinently, it has allowed the calculation of the affinities of the STAT1 SH2 domain for a range of phosphopeptides, including that derived from the IFNγ receptor (the Y440 described above[314] and Lackmann, Harpur, Oates and Wilks, unpublished). This binding is high affinity, and is sensitive to the precise amino acid context in which the phosphotyrosine is embedded.[314] As predicted, STAT2 does not bind this phosphotyrosine motif (Lackmann, Harpur, Oates and Wilks, unpublished). Furthermore, STAT2 is able to bind to a specific phosphotyrosine motif from the IFNα/β receptor (Y466),[115,314] thus an important level of specificity has been revealed, by means of which the precise composition of the STAT transcription factor complex would be determined.

TYROSINE PHOSPHORYLATION OF STATs

Having been recruited to the activated cytokine receptor by means of the receptor-borne phosphotyrosine residues, the relevant STAT (or STATs) becomes available as a substrate for the activated JAK. As described above, the STATs possess a single conserved tyrosine phosphorylation site, It is now clear that this tyrosine has a major role to play in the coordination of the

STAT-dependent signal transduction pathways which lie downstream of cytokine receptors. Thus, having been summoned to the receptor by means of the receptor phosphotyrosine residues, the STAT's resultant proximity to the JAK leads to a phosphorylation of the STAT on tyrosine. A single tyrosine residue located in the C-terminal portion of the STAT protein (amino acid position 701 in STAT1 (Y701)[240] or the analogous position of the other STAT family members) becomes phosphorylated. While this evidence was tantalizing enough to suggest the STATs were substrates for the JAKs, it remained to be demonstrated biochemically. A number of independent studies have demonstrated, with the aid of synthetic peptides or co-expression of JAKs and STATs in insect cells,[387] that the STAT C-terminal tyrosine residues are indeed substrates for active JAK enzymes. Moreover, the phosphorylation of these tyrosine residues was both necessary and sufficient to lead to the generation of activated DNA-binding complexes in the insect cell extracts. Interestingly, in this study, JAK1, JAK2 and Tyk2 were all equally capable of tyrosine phosphorylating STAT1, suggesting that the selection of a particular JAK family member by a cytokine receptor does not appear to be a critical parameter governing the specificity of signal response in vivo.

JAKs and STATs Act Downstream of Receptor Tyrosine Kinases

Activation of STAT-containing DNA binding complexes by RTKs followed closely upon the discovery that these complexes were recruited by cytokine receptors. For example, EGF,[197,250,278,280] CSF-1[388] and PDGF[256,258] have all been shown to induce the formation of STAT containing transcription factor complexes. For the cytokine receptors, the recruitment of members of the JAK family of PTKs fulfills an obvious inadequacy; without them the tyrosine kinase activity required for cytokine dependent signal transduction would not be available. In the case of the RTKs, however, there is already a PTK domain in place; is there a role for the JAK family of kinase here, or is the intrinsic PTK activity of this class of receptors all that is required? Preliminary studies in mouse 3T3 cells suggested that JAK1, JAK2 and Tyk2 all became phosphorylated upon tyrosine in response to the growth factor

PDGF.[389] Furthermore, all of these cytoplasmic PTKs appeared to be associated with the activated PDGFR-β. In order to establish whether JAK1, JAK2 or Tyk2 were required for the generation of SIE binding STAT complexes, the human fibroblast cell lines used in the genetic dissection of the IFN pathway[285,289,295,299,300] (see chapter 6) were infected with a retrovirus bearing the PDGFR-β. After establishing that levels of PDGFR-β were equal in these cell lines, it was found that the absence of neither JAK1, JAK2 or Tyk2 made any difference to the generation of STFs by PDGF. Moreover, in each of the JAK-deficient cell lines, both of the remaining functional JAKs became phosphorylated upon PDGF treatment of the cells, suggesting that, unlike the IFN signaling pathway, there is no obligate interdependence of these PTKs in their response to PDGF. This observation might suggest one of two possible interpretations; firstly, the recruitment and phosphorylation of members of the JAK family of PTKs is an epiphenomenon in PDGF signaling; or, alternatively the JAKs operate interchangeably so that the absence of any one of them is insufficient to lead to the breakdown of PDGF signaling.

JAKs and STATs Act Downstream of G-Protein Coupled Receptors

The activation of JAKs and STATs also appears to be a feature of G-protein coupled receptors such as the angiotensin II AT_1 receptor.[390-392] Angiotensin II is the effector molecule of the renin angiotensin system, and it regulates a range of important hemodynamic control mechanisms, such as vasoconstriction and aldosterone release. The AT_1 receptor is a member of the G-protein coupled superfamily. The demonstration that smooth muscle cells exposed to angiotensin II resulted in the association of JAK2 with the AT_1 receptor, the tyrosine phosphorylation of JAK2 and the recruitment of STAT1 and STAT2, was thus a surprise, and suggested that this class of receptors may also plug into the JAK-STAT pathway in order to achieve the many and varied physiological consequences that angiotensin II can bring about. It remains to be demonstrated that this recruitment of the JAK-STAT pathway is a general feature of signaling in this class of receptors.

DIMERIZATION AND ACTIVATION OF STATs

The dimerization of the STATs, both homodimerization (for example, that initiated with STAT1 by IFNγ stimulation[238]) and heterodimerization (for example, that initiated for STAT1 and STAT2 by IFNα/β stimulation[203]), is mediated by the interaction of the 701 (or its equivalent) phosphotyrosine with the SH2 domain of its opposite number in the dimer, whereupon the STAT is able to dimerize in a head-to-tail conformation with its partner (see Figs. 20 and 21). This conclusion was drawn from comparisons with results derived from the publication of the X-ray crystal structure of the SH2-SH3 domains of the *src* family kinase *LCK*,[33] whereby a head-to-tail configuration was adopted by the dimerizing proteins. Here, the phosphotyrosine of one *LCK* is bound by the SH2 domain on its binding partner, and vice versa. Interestingly, in the case of *LCK*, this head-to-tail configuration requires the interaction of the SH3 domain with structures within the SH2 domain of the complementary protein to stabilize the complex.

The most likely order of events then, is the recruitment of the appropriate STAT to the phosphotyrosine decorated intracellular domain of the activated cytokine receptor, followed by the JAK dependent phosphorylation of the C-terminal tyrosine (Y701). What are the motive forces which drive the monomer STAT from the receptor into a dimer? What defines the precise composition of the STAT dimers? The availability of real-time measurement of the kinetics of interaction of STAT SH2 domains and their phosphopeptide ligands has permitted the resolution of some of these unresolved mechanistic questions.

Peptide competition experiments on the BIAcore™ demonstrated that binding of purified STAT1 to the IFNγ-receptor phosphopeptide Y440 was inhibited to varying degrees by tyrosine phosphopeptides corresponding to the C-terminal tyrosine residue on each of the STAT proteins. The inhibitory activity (ID_{50} value) of each phosphopeptide was as follows:[314]

Peptide	ID_{50}
IFNγR tyrosine 440	0.55 μM
STAT1 tyrosine 701	60 μM
STAT2 tyrosine 690	7.5 μM

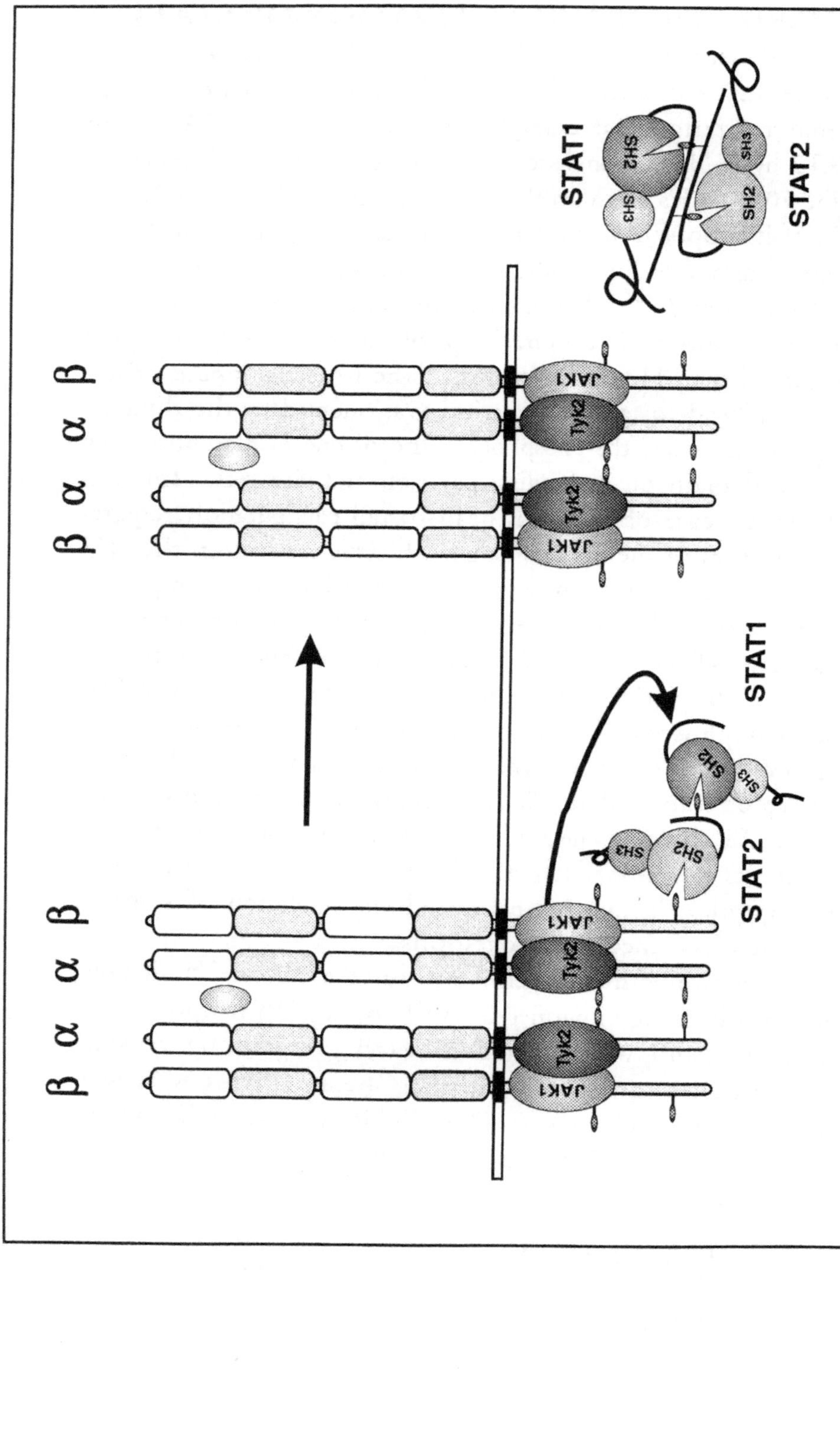

Fig. 20. Dimerization and activation of STATs. The tyrosine phosphorylated STAT2 provides a docking site for STAT1, which proceeds to bind to the receptor bound STAT2. The STAT1 becomes a substrate for the activated PTK activity of JAK1 and Tyk2, and the STATs form a heterodimer held together by the mutual contact of the phosphotyrosine residue of one STAT with the phosphotyrosine of the other.

STAT3 tyrosine 705	33 μM
STAT4 tyrosine 694	80 μM
STAT5 tyrosine 694	300 μM
STAT6 tyrosine 641	32 μM

Thus, these figures appear to suggest that while the STAT1 SH2 domain does interact with the phosphotyrosine in the C-terminal tail of the complimentary STAT1, it does so with a far lower affinity (ID_{50} = 60 μM) than that observed for IFNγR tyrosine 440 phosphopeptide (ID_{50} = 0.55 μM) or the STAT2 C-terminal phosphopeptide (ID_{50} = 7.5 μM).[314] This hierarchy of affinities implies that there must be an underlying mechanism other than the higher affinity for the C-terminal phosphotyrosine on another STAT1 molecule, driving the phosphorylated STAT1 from the IFNγR α chain. While this question is not yet resolved, it seems likely that the head-to-tail conformation of the two STAT monomers assembled in the activated STAT dimer confers a higher affinity state than that generated by the STAT/receptor complex. Once formed, the STAT dimer is thus driven from the receptor, and moves to the nucleus.

SERINE PHOSPHORYLATION OF STATs

While the tyrosine phosphorylation of STAT family members has been linked mechanistically to their activation as transcription factors for some time now (see above), the demonstration that serine phosphorylation of STATs is also a feature of their recruitment into active transcription complexes is of more recent origin[241,249,254,393] and remains obscure in respect to its function (see chapter 4, section entitled, "The S727 Phosphorylation Site"). The first demonstration that serine phosphorylation of STATs was an important feature of signal transduction in the JAK-STAT pathway emerged from the use of the serine/threonine kinase inhibitor.[254] Treatment of cells with this inhibitor resulted in the inability of IL-6 induced STAT3 homodimers to form functional DNA binding complexes. Interestingly, phosphatase treatment of "activated" STAT3 resulted in abolition of binding to a Ly-6E GAS element (a low affinity GAS element, see chapter 6), although binding to other high affinity GAS elements was only slightly reduced.

One possible interpretation of these data is that the stability of the STAT3/STAT3 homodimers, but not STAT1/STAT3 heterodimers was compromised by the inhibition of serine/threonine kinase activity within the cell. Alternatively, recruitment of ancillary factors to the transcription factor complex may be serine phosphorylation dependent, resulting in the loss of the capacity to form a complete STAT3-containing transcription factor complex. Importantly, this work brought the JAK-STAT pathway into juxtaposition with the MAP kinase pathway, an area that will be discussed in chapter 9 of this book (see section entitled "Integrating the JAK-STAT Pathway into the Cellular Signaling Network").

In the case of STAT1α activation following IFNγ treatment of cells, serine phosphorylation also appears to be an important step in the process.[241] The demonstration that the site of Serine phosphorylation on STAT1α was Serine 727 (a sequence conforming to the MAP kinase substrate consensus) has provided no real clues as to the functional basis for this modification, however, it is worth noting that the STAT1β isoform lacks this phosphorylation site,[202,236] and is as a consequence unable to activate some of the genes possessing STAT1 binding sites, such as IRF-1.[241] Thus, although clearly important to the signaling process, the role of serine phosphorylation of STATs in the JAK-STAT pathway remains enigmatic.

NUCLEAR TRANSLOCATION OF STATs AND ACTIVATION OF GENES BY STAT-CONTAINING DNA BINDING COMPLEXES

STATs Activate Gene Expression in a Promoter-Specific Fashion

The characterization of DNA binding elements for the STATs reveals a rather complex situation, with a range of DNA elements based around a palindromic 10-12 base pair consensus sequence (-TTTNNXN'N'AAA-), each exhibiting varying degrees of affinity for the STAT complex. DNA consensus sequence elements have been elucidated for IFNα/β (ISREs; reviewed in ref. 221), IFNγ (GAS; reviewed in ref. 221), IL-6[394] and IL-4[230] dependent genes among others, and EMSA experiments have revealed that many of

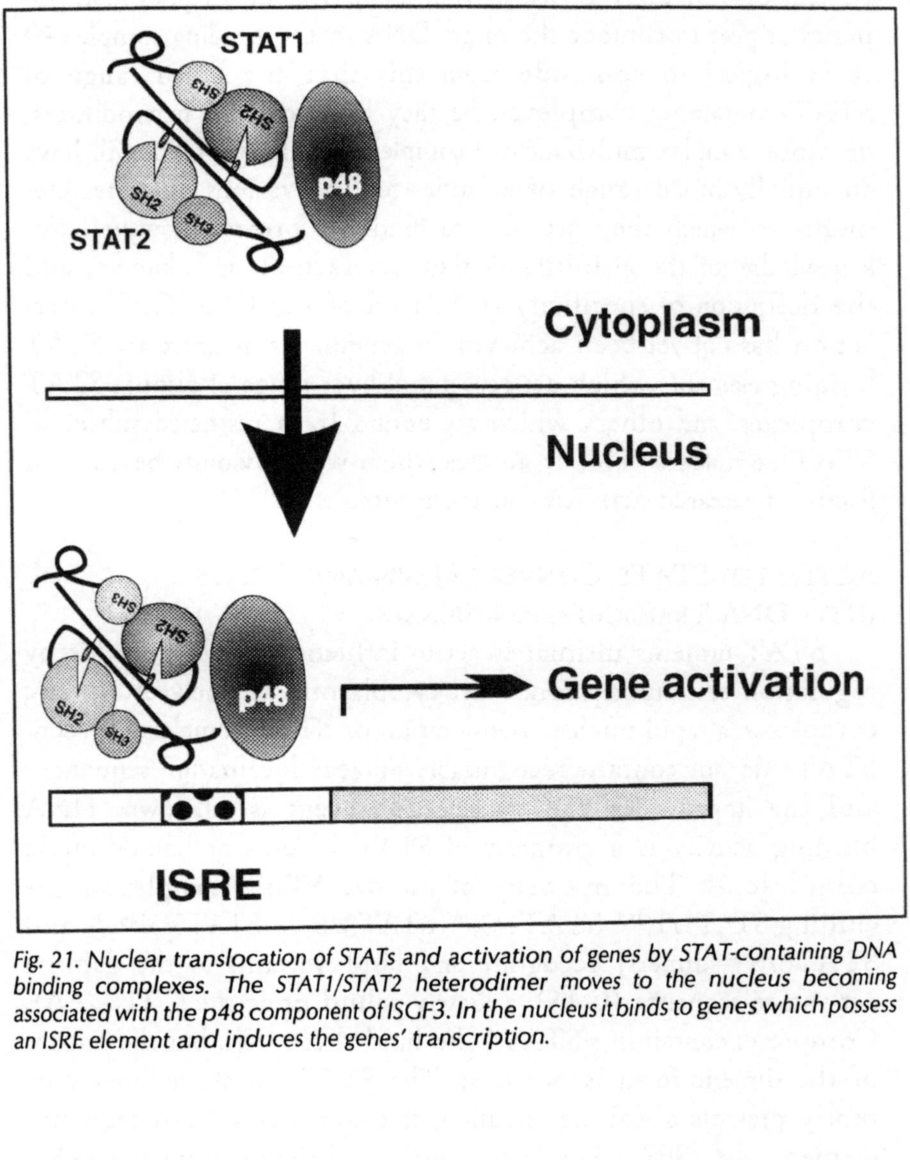

Fig. 21. Nuclear translocation of STATs and activation of genes by STAT-containing DNA binding complexes. The STAT1/STAT2 heterodimer moves to the nucleus becoming associated with the p48 component of ISGF3. In the nucleus it binds to genes which possess an ISRE element and induces the genes' transcription.

these elements are targets for STAT binding (reviewed in ref. 221). In the case of ISGF3 (see above), all of the polypeptide components appear to contact the target DNA in the binding complex.[265] It is logical to conclude from this that the broad range of STAT-containing complexes, be they homodimers, heterodimers, or more complex multifactorial complexes such as ISGF3, will have an equally broad range of affinities for the various promoter elements to which they are able to bind. At present, however, our knowledge of the affinities of these interactions is unknown, and the definition of specificity at the level of the DNA/STAT interaction has not yet been achieved. In general terms, there are STAT binding elements which are recognized by a range of distinct STAT complexes, and others which are bound by a restricted subset of STAT complexes. This is an area which will obviously be a major focus of research activities in the future.

Activated STATs Convert Membrane Events into DNA Transcription Signals

STAT proteins ultimately act to influence a cell's behavior by regulation of transcription. After cytoplasmic assembly of dimeric complexes, a rapid nuclear translocation of STAT complexes is seen. STATs do not contain recognizable nuclear localization sequences and the impetus for the translocation event is unknown. DNA binding activity is a property of STAT hetero- or homodimeric complexes.[238] The majority of known STAT complexes, including STAT1/1,[238] STAT1/3,[206] STAT3/3,[206] STAT5/5[146,395] and STAT6/6[264] dimers, recognize and bind variants of the gamma activation sequence (GAS), a loosely palindromic site ($TTN_{5-6}AA$). Complexes containing STAT4 also bind these sites, but the nature of the dimeric form is not clear. The STAT1/2 heterodimer currently presents a unique situation; it recognizes a DNA sequence element, the ISRE, that is substantially different from the GAS, and it requires the presence of the interferon consensus sequence binding protein (ICSBP) family member ISGF3γ (p48) as a cofactor for high affinity DNA binding.[220,265,396]

Experiments designed to select the optimal binding site for a homodimer of STAT1 or STAT3 in vitro reveal an almost indistinguishable choice by both STATs ($TTCC^{G}/_{C}GGAA$).[242] However, STAT1 and STAT3 binding specificities to GAS sites from natural SBEs vary considerably, suggesting that in vivo, a range of suboptimal sites allow selective transcriptional responses.[242] Another form of selectivity is demonstrated by determination of the optimal binding site of the STAT6 homodimer, which shows a preference for an increased spacing between the palindromic repeats TTC and GAA of four nucleotides, compared to the STAT1 optimal spacing of three nucleotides.[230] The optimal binding sites of STAT4 or STAT5 homodimers have not been measured.

Single minimal GAS elements appear to function poorly in the activation of heterologous genes, but multiple copies arrayed in tandem are highly efficient.[397] Indeed, SBEs are not found in isolation in the promoters of responsive genes. Although electrophoretic mobility shift assays (EMSA) using minimal GAS elements provide a simple and accurate method for determining which STAT proteins are activated as a result of a signaling event, they do not reveal the complexity of interactions at a promoter in situ. Three examples drawn from IFNγ signaling serve to illustrate this notion. IFNγ induces the formation of GAF, the STAT1 homodimer, upon minimal GAS elements derived from many IFNγ inducible genes. In contrast, at the extended interferon response (IR) site in the IRF-1 gene promoter, after IFNγ engagement of the signaling system, three complexes known as the IRBFγ form, all of which contain STAT1.[398,399] Since only one of these complexes can correspond to the classic IFNγ induced STAT1 homodimer, the others must contain additional factors.

Additionally, IFNγ induces the formation of γRF-1 upon the tandem GAS containing 41 base-pairs γRE-1.[400] The IFNα/β complex is distinct from GAF in that it is induced very rapidly, is not sensitive to cyclohexamide treatment, and displays a slower electrophoretic mobility. Furthermore, γRE-1 affinity purified γRF-1 cannot bind the GBP-derived GAS element, but despite this, it can be supershifted with antisera to STAT1 (but not STAT2, 3 or 4), demonstrating that novel STAT1 containing complexes are formed from the IFNγ-R.

Finally, the gamma response region (GRR) which confers IFNγ inducibility to the FcγR gene, has an extended structure over 35 base pairs, consisting of a GAS, sufficient for minimal transcriptional activation, but including additional sequence necessary for a full transcriptional response.[401,402] Upon IFNγ stimulation up to three distinct complexes termed FcRF form upon the GRR, depending on cell type, two of which contain STAT1.[401] The more rapidly migrating complex of the two forms on the GRR-derived GAS alone, and is presumably a STAT1 homodimer (GAF), whereas the slowest migrating complex depends upon the additional sequence in conjunction with the GAS and must contain factors additional to STAT1.[402] The nature of these components is unknown; they may be other STATs, or unrelated proteins, by analogy with the presence of ISGF-3γ (p48) in the ISGF-3/ISRE interaction, or they may be variously post-translationally modified.

Thus, it is tempting to speculate that the multimeric ISGF3 complex is a better model for STAT promoter interaction in general. Clearly, the core notion of STAT1 homodimer-GAS element interaction that is derived from selection for disruption of IFNγ responsive promoter induction is a simplification, and may dramatically underestimate the range of STAT-containing complexes that are involved in transcriptional regulation in vivo.

TYROSINE PHOSPHATASE MEDIATED RESETTING OF THE PATHWAY

The signals transmitted by the circuitry discussed above are, by necessity, transient in nature. In order to reset a signaling pathway that is largely coordinated by phosphotyrosine/SH2 domain interactions, it seems plausible that a phosphotyrosine phosphatase would be required to extinguish the signal initiated by the PTK activity of the relevant JAK. Three particular substrates must be dephosphorylated: the activated JAK; the receptor borne SH2 binding site(s); and the phosphotyrosine residue involved in the STAT dimer formation. Recent work in the EPO-R system appears to have brought forward a candidate for the first two of these processes.[403] In an important series of experiments, it was demonstrated that the binding of SH-PTP1 (a member of the sub-family of phosphotyrosine phosphatases [PTPases] which possess SH2 domains) to the tyrosine phosphorylated form of EPO-R resulted in the activation of its catalytic activity, which in turn resulted in

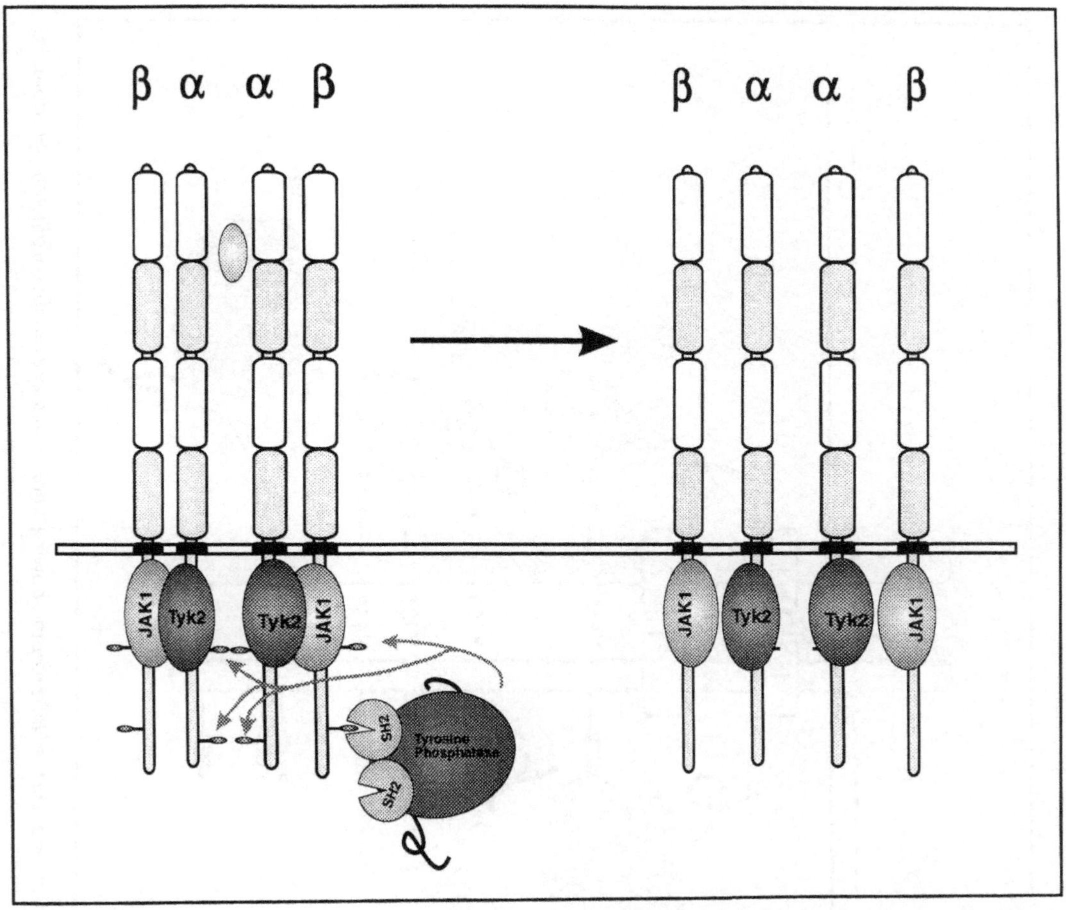

Fig. 22. Tyrosine phosphatase mediated resetting of the pathway. The tyrosine phosphatase SH-PTP associates with the phosphotyrosine decorated intracellular domain of the IFNα/β receptor. The action of the phosphatase enzyme removes the phosphotyrosine residues and resets the pathway.

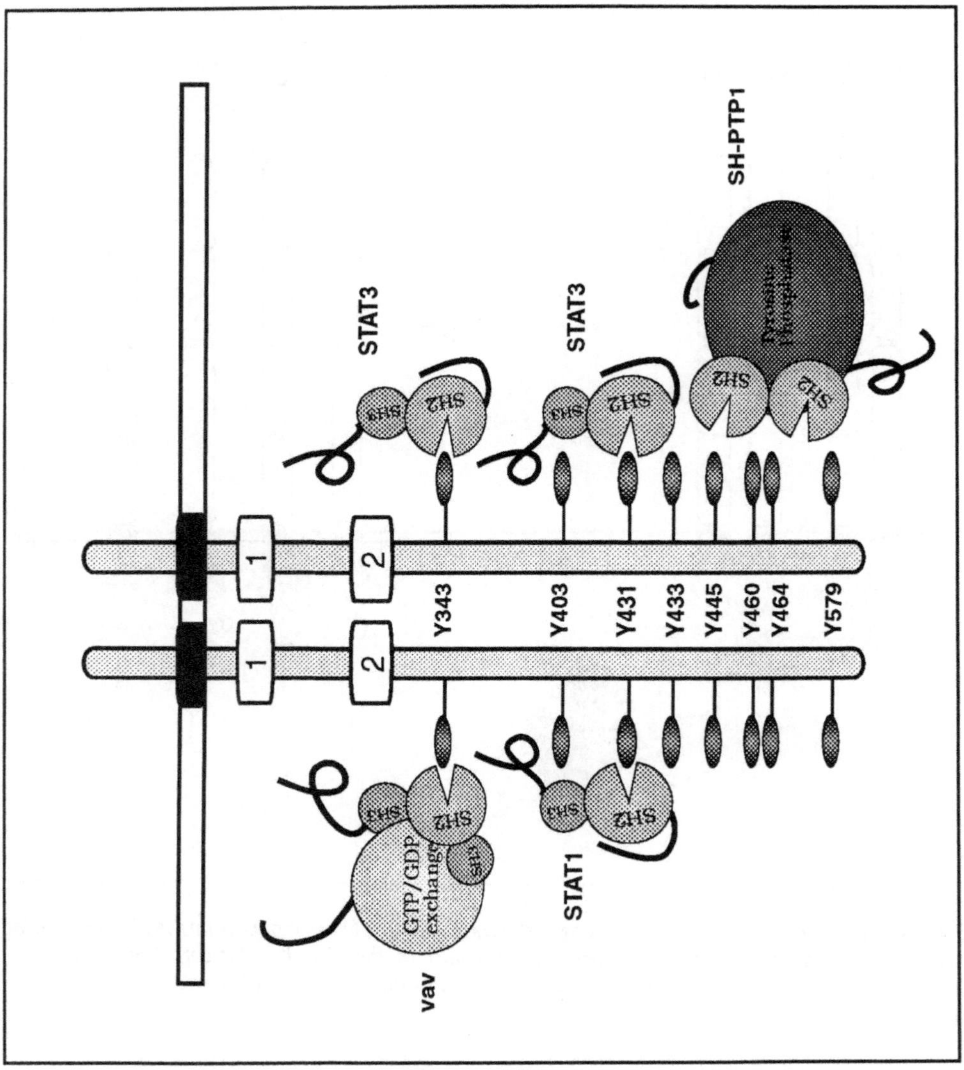

Fig. 23. Intracellular domain of the EPO receptor showing binding sites for a number of different SH2 domains.

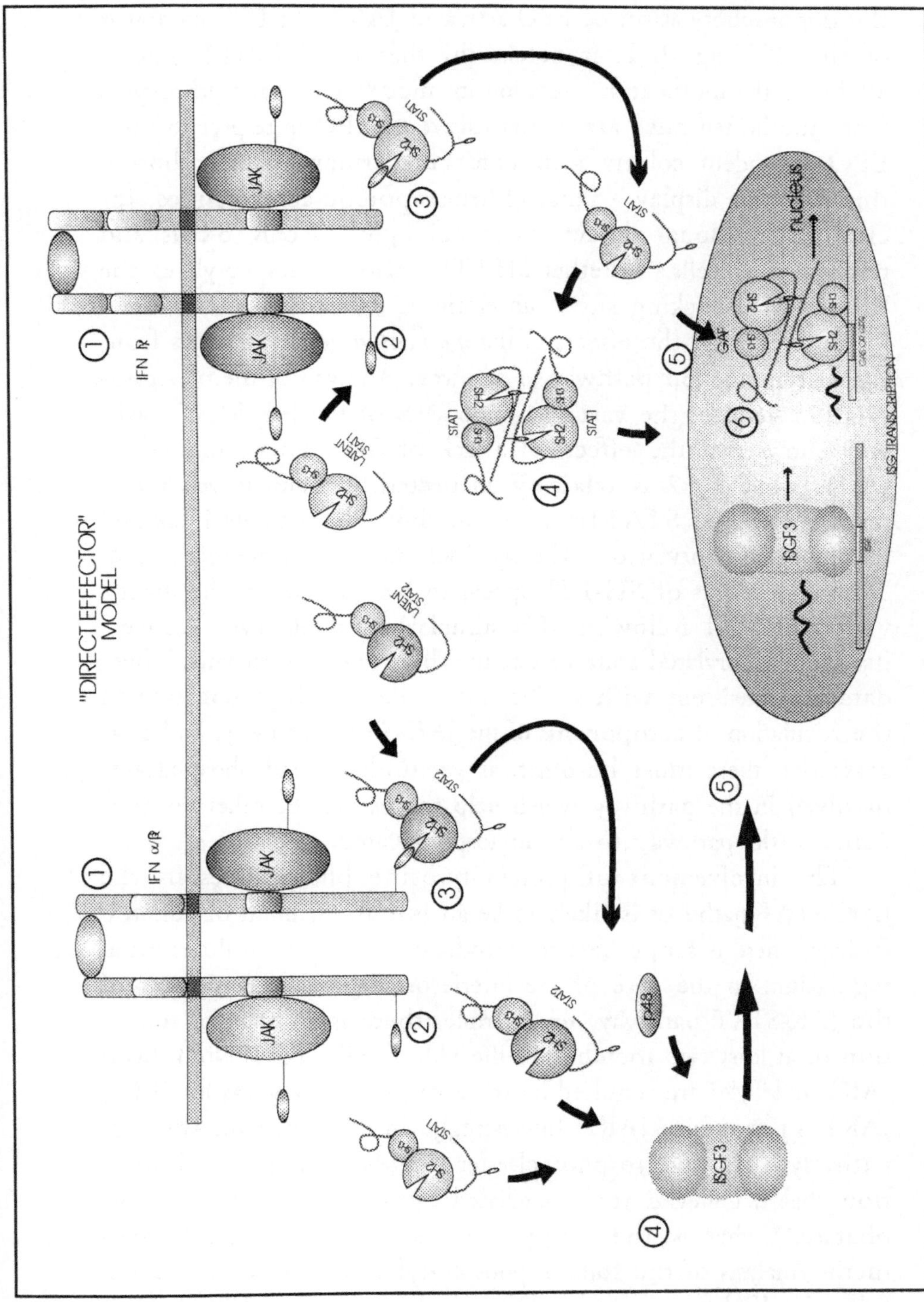

Fig. 24. Global view of the JAK-STAT pathway using both IFNα/β and IFNγ as examples.

the dephosphorylation of EPO activated JAK2, the PTK mediator of the EPO signal. It is noteworthy that the SH-PTP1 gene is allelic to the motheaten mutation in mice[404] and that erythrocytes from motheaten mice are hypersensitive to EPO in respect to their EPO-dependent colony formation. Furthermore, mice harboring this mutation display a range of hematopoietic abnormalities, including functional defects in granulocytes, T cells, B cells and natural killer cells. Whether SH-PTP1 also dephosphorylates the EPO-R SH2 docking site is uncertain at present.

Interestingly, the effects of the *motheaten* genotype varies from signal transduction pathway to pathway. Analysis of the phosphorylation status of the various components of the JAK-STAT pathway shows that the effects of a lack of SH-PTP are limited to JAK1, while Tyk2 is relatively unaffected.[405] Thus, in *motheaten* mice, JAK1 and STAT1α appear to show dramatically increased levels of phosphorylation, whereas Tyk2 and STAT2 are unaffected. Thus, the effects of SH-PTP appear to be restricted to the inactivation of JAK1 following IFN stimulation, while Tyk2 recovers its unphosphorylated state despite the absence of this enzyme. These data are consistent with a differential role of this phosphatase in the regulation of components of the JAK-STAT pathway, and suggests that there must be other, as yet undiscovered phosphatases involved in the pathway which help to restore the other components of the pathway to an unphosphorylated state.

The involvement of protein tyrosine phosphatases in the JAK/STAT pathway is likely to be an important point of control. Indeed, there is scope here to introduce an element of differential regulation. In the case of the interferon-dependent activation of the JAK-STAT pathway, for example, there is an obligate activation of at least two members of the JAK family. For IFNα/β, both JAK1 and Tyk2 are required for correct signaling, whereas for IFNγ, JAK1 is paired with JAK2. Interestingly, in monocytes the IFNα/β, pathway is sensitive to phorbol ester (PMA) treatment, an inhibition that is believed to be mediated by a protein tyrosine phosphatase,[406] whereas, the IFNγ pathway is insensitive to PMA treatment. Analysis of the state of phosphorylation of JAK1 and Tyk2 following IFN treatment, in the presence or absence of PMA, suggested that this difference in sensitivity is due to a differential dephosphorylation of Tyk2 and JAK1 by a PMA induced protein tyrosine phosphatase. Since Tyk2 is only recruited by the IFNα/β system, whereas JAK1 is involved in both IFNγ and IFNα/β signaling, it seems likely that the differential sensitivity of these two signaling pathways to PMA is due to a selective inactivation of the Tyk2 activity.[406] It seems likely therefore, that the regulation of the JAK-STAT pathway may occur at least two levels, namely, initiation of signaling (by PTKs) and specific attenuation or inactivation of signaling by protein tyrosine phosphatases. Moreover, there is now evidence that there is at least one other protein tyrosine phosphatase involved in the inactivation of the components of an activated JAK-STAT pathway.

Currently, the most difficult aspect of the attenuation of the JAK-STAT pathway to explain is the removal of activated STAT dimer from the system. Whether this is brought about by a nuclear phosphatase, or by some other degradative mechanism is completely unknown and remains to be determined.

DISCUSSION

The foregoing discussion of the essential elements of the JAK-STAT pathway has been written against the tide of rapidly changing ideas based upon a constant influx of new information. The fundamentals are unlikely to change too much in the foreseeable future; a JAK family member (or members) will remain the key tyrosine kinase that becomes activated in response to cytokine binding; the tyrosine phosphorylated intracellular domain of the cytokine receptor will still be the major site of recruitment of STATs and other signaling molecules; and tyrosine phosphorylation mediated dimerization of the STATs will still be the pivotal event in the formation of activated STAT-containing transcription complexes. Nonetheless, this is a swiftly moving area, and new refinements of the basic principles are being uncovered at a rapid rate. A number of important new approaches are being undertaken in the detailed analysis of this pathway. We have dedicated the final two chapters of this book to an analysis of their impact. Thus, in the next chapter, the use of genetically tractable organisms as a tool to examine the JAK-STAT pathway will be discussed, and in the final chapter, we will sample a range of the most important future trends in research into the JAK-STAT pathway. What will emerge from these studies will no doubt still be recognizable as the pathway that we have just described. Nonetheless, the integration of this pathway into the overall signaling network of the cell, and the explanation of the driving forces which coordinate the molecular choreography of this pathway, will be a task that which will take many years to complete.

CHAPTER 8

Hopscotch/Tumorous Lethal: A JAK Family Member from *Drosophila melanogaster*

Historically, organisms such as the fruit fly (*Drosophila melanogaster* or *D. melanogaster*) and the nematode worm (*Caenorhabditis elegans* or *C. elegans*), for which sophisticated genetic approaches have been developed, have proven to be immensely useful tools for the investigation of signal transduction pathways. The degree of identity which underlies the intracellular signal transduction pathways of organisms as diverse as *C. elegans* and the laboratory mouse (*Mus musculus)* is reassuring, and suggests that the information which has emerged from the study of lower metazoans is tremendously pertinent to our notions of how signal transduction works in human cells. In choosing to explore a signal transduction pathway such as the JAK-STAT pathway in mammalian cells, the limitations of the experimental systems available must be acknowledged. Thus, despite the remarkable advances in our understanding of this pathway which have emerged from the study of cellular mutants of interferon-dependent signaling, coupled with the intense focus on the biochemistry of this pathway in a host of cytokine-dependent signaling pathways, there remain limitations to each of these approaches which will confound a fuller understanding of how the JAK-STAT pathway fits into the network of signal transduction within any given cell. For example, investigation of the interferon-dependent signaling mutants described in chapter 6, will only facilitate an examination of the JAK-STAT pathway in the context of the activation of the particular promoter used in the selection protocol under which the cells were originally generated. In contrast, the use of the coordinated embryonic development of an entire organism as an "assay" for molecules which might interact with components of the JAK-STAT pathway(s) is likely to present a more realistic picture of how it is integrated into the intracellular signaling network of many thousands of developmentally different cells and cell types. Thus, while such an assay system is more difficult to establish, the dividends of success would be considerably greater.

In order that an organism-based genetic screen might be established, several key elements need to be in place. Firstly, it must be believed that signal transduction works in essentially the same way in lower eukaryotic cells as in those from higher eukaryotes. At present, it is widely accepted that the broad sweep of biochemical underpinnings of signal transduction is essentially identical in all metazoan organisms. Secondly, a homologue of the molecule of interest must be present in the experimental organism to be employed. Thirdly, a genetic screen must be made possible, either via the use of a dominant allele of the gene of interest, or through the availability of a null mutant of the gene of interest from which a useful phenotype might emerge, and with which one might attempt to establish epistasis of the gene of interest with respect to genes defective in other mutants with similar phenotypes.

Several elegant studies have been undertaken along these lines with highly successful screens for genes that disrupt neuronal differentiation in the *Drosophila melanogaster* compound eye,[407,408] and define vulval cell fate in the nematode *Caenorhabditis elegans.*[409-411] The genes of the PTK-ras-MAPK pathway have been recovered in each of these screens and have revealed these pathways to be completely homologous to the signal transduction mechanisms which lie downstream of the EGF-R in mammalian cells. This high level of conservation of mechanism, despite the diverse nature of the biological signal across very distant taxa, implies that genetic screens such as these may be of particular use in the exploration of other signal transduction pathways.

The discovery of a *Drosophila melanogaster* JAK family member (*hopscotch*) was not unexpected. However, that it was allelic to a gene important for embryonic patterning was largely unanticipated. The discovery of a hypermorphic mutant of the same gene product (known as *Tumorous lethal* or *Tum*^{l}) produced an unexpected bonus and has greatly increased the prospects of useful information emerging from the study of the JAK-STAT pathway in *Drosophila melanogaster.* We will outline the *hopscotch/Tum*^{l} story below. However, in order to understand the implications of the *hopscotch/Tum*^{l} phenotype, a few brief words on the subject of *Drosophila melanogaster* segmentation are required.

DROSOPHILA MELANOGASTER EMBRYOGENESIS

The basic segmental body pattern of *Drosophila melanogaster* is established during embryogenesis. The *Drosophila* embryo develops from a fertilized egg through to a hatched larva in 24 hours. The first stage of embryogenesis following fertilization is the generation of a multinuclear syncytial stage (known as the syncytial blastoderm) developed by dint of 13 rounds of mitosis in the absence of cellularization. During this stage, the embryo's developmental program runs off a store of maternal mRNA laid down during the process of oogenesis. At the end of the 13th mitosis, after two hours of rapid nuclear division, the nuclei migrate towards the periplasm, the zygotic genome begins to be transcribed, and the embryo begins to cellularize (a stage known as cellular blastoderm). Thereafter, the process of segmentation begins, followed rapidly by gastrulation and germ-band elongation.

Early segmentation events come about as a consequence of the expression of a number of transcription factors which are expressed within the syncytial blastoderm.[412-414] The first stage of this process is the deposition of transcripts of the *bicoid* (*bcd*) gene at the anterior end of the egg, where they are transcribed to generate a morphogenetic gradient of *bcd* protein.[415,416] After the establishment of the anterior-posterior axis in this fashion, a chain reaction of regulated expression of two classes of transcription factors begins. These transcription factors are expressed in a coordinated fashion within the embryo, and define domains of the blastocyst which will ultimately resolve into the cuticular segments of the larva. The first wave of zygotically expressed segmentation genes are known as the Gap genes[412,414,417] and include genes such as *hunchback* (*hb*), *giant* (*gt*), *Krüppel* (*kr*), and *knirps* (*kni*). The expression pattern of these genes divides the embryo into broad anterior-posterior segments. These broad segments are refined by means of the expression of a second wave of transcription factors known as the Pair-rule genes (examples of which rejoice under names such as *runt* (*run*), *fushi tarazu* (*ftz*), and *even-skipped* (*eve*)). The Pair-rule genes are expressed in seven stripes, each of which is about four nuclei wide, and the boundaries of which are defined by the activity of the proteins encoded by the Gap genes.

Loss-of-function mutants have been discovered for each of the Gap and Pair-rule genes and their necessity for the process of segmentation is clearly demonstrated from the mutant phenotypes. Significantly, all of the Gap and Pair-rule genes so far uncovered are transcription factors, and the exquisite interplay of their expression patterns appears to be due to the presence of binding sites for other members of the family upstream of each gene encoding a segmentation gene. Thus, the genes regulate each other's expression and force the broader expression pattern exhibited by each gene to resolve by means of mutual regulation of each other. Stripe-specific expression of Pair-rule genes is mediated by transcriptional enhancers, capable of binding maternal and GAP gene products. For example, a 5.2 kb piece of DNA from the proximal *even-skipped* (*eve)* promoter drives the expression of β-galactosidase (forthwith known as *eve*5.2-*lac*Z) in the same pattern as *eve* protein namely, in stripes 2, 3 and 7. Interestingly, individual stripe-specific enhancers have been defined for a number of Pair-rule. For example, a 500 bp region from the original 5.2 kb *eve* promoter described above, will cause a *lac*Z reporter gene to be expressed in an identical position to *eve* stripe 3, suggesting that this region contains a stripe-specific promoter.[177]

The hierarchical relationship between the members of the Gap and Pair-rule genes ensures the correct segmentation patterning of the *Drosophila* embryo, and provides a number of interesting experimental opportunities for the investigation of the developmental control of this process. That these segmentation genes were all transcription factors suggested an elegant, albeit relatively oversimplified, mechanism for the developmental regulation of this process. The discovery that a *melanogaster* JAK family member was critically involved in embryonic patterning was thus an intriguing one, and has been important in refocussing our attention onto the broader range of regulatory processes which characterize embryonic patterning.

DISCOVERY OF *HOPSCOTCH*

While it is true that most of what we now know about the JAK-STAT pathway has come from a remarkable synergy between biochemical characterization of a number of "interesting" proteins, and a somatic cell genetic approach aimed at uncovering the components of the IFN signal transduction pathway, with the (inevitable) discovery of a *melanogaster* JAK family member, a host of previously unanticipated discoveries may well now be within reach. While the somatic cell genetic approach to the IFN signal transduction pathway has indeed provided us with a paradigm shift in our understanding of the events which lie downstream of cytokine receptors, these experiments, perforce, allow only a glimpse at those events which are central to the process of the activation of a particular gene by a particular cytokine. The opportunity to examine the events which take place at the level of the development of an entire organism means that many other pathways which converge with the pathway under direct examination, or indeed those which diverge from it, are all available for investigation. The highly sophisticated nature of our understanding of the genetics of development in *D. melanogaster* makes it a system with enormous potential for the discovery of the unexpected.

hopscotch (*hop*) emerged from a screen for X-linked maternal effect zygotic lethal loci.[418] *hopscotch* flies die in the late larval or early pupal stage, and *hopscotch* can thus be characterized as an X-linked larval/pupal zygotic lethal mutation. Seventeen distinct alleles of *hopscotch* have been uncovered,[418] ranging in phenotype from mild defects in segmentation (loss of one abdominal segment; A5) to severe effects on segmentation (defects in five segments; two thoracic segments, T3 and T4; and three abdominal segments, A4, A5 and A8; see Fig. 26D). All alleles of *hopscotch* mapped to chromosomal bands 10B6-8 (meiotic position 34.61) on the X-chromosome.[418] Interestingly, trans-allelic complementation between each of the *hopscotch* alleles suggested that there are two classes of amorphic *hopscotch* mutants (class A and class B),[418] and a hypomorphic mutant known as $l(1)hop^{msv1}$. Heterozygotes produced from crossing the hypomorphic mutant $l(1)hop^{msv1}$ and any of the nine class A mutants were inviable, whereas, crossing $l(1)hop^{msv1}$ with any of the seven class B mutants resulted in between 12% (for one mutant) and 100% complementation (for three mutants). Thus, there appeared to be two distinct functions emerging from the *hopscotch* gene product. The full importance of these observations is yet to be understood.

The *hopscotch* gene was cloned and characterized in the early 1990s and published in 1994.[177] The *hopscotch* gene is composed of ten exons, and covers approximately 7.5 kb. Northern blot analysis of all of the developmental stages of *Drosophila* embryogenesis with a *hop* specific probe revealed a maternal mRNA of 5.1 kb, and a zygotic transcript of 5.4 kb, while in situ hybridization revealed a uniform pattern of expression from blastoderm to germ band extension. The deduced amino acid sequence suggested that the *hop* locus encoded a PTK. Subsequently, sequence comparisons between the deduced Hop protein and other PTKs revealed that it was a member of the JAK family.[166,167,180] Hop shares the same basic architecture characteristic of this family of PTKs, in particular the presence of a second kinase-related domain N-terminal of the PTK domain. The PTK domain and the JH2 domains of Hop and JAK1 are 39% and 27% identical, respectively. Comparison of the four mammalian members of the JAK family revealed the presence of five further domains held in common. Extension of this analysis to include Hop indicated that this extensive homology across the length of the protein is not well conserved within the Hop protein with only five of the JH domains being recognizable in Hop (Fig. 25).

At present, there have been very few biochemical studies undertaken upon the Hop protein, although it has been demonstrated that this protein possesses PTK activity.[419]

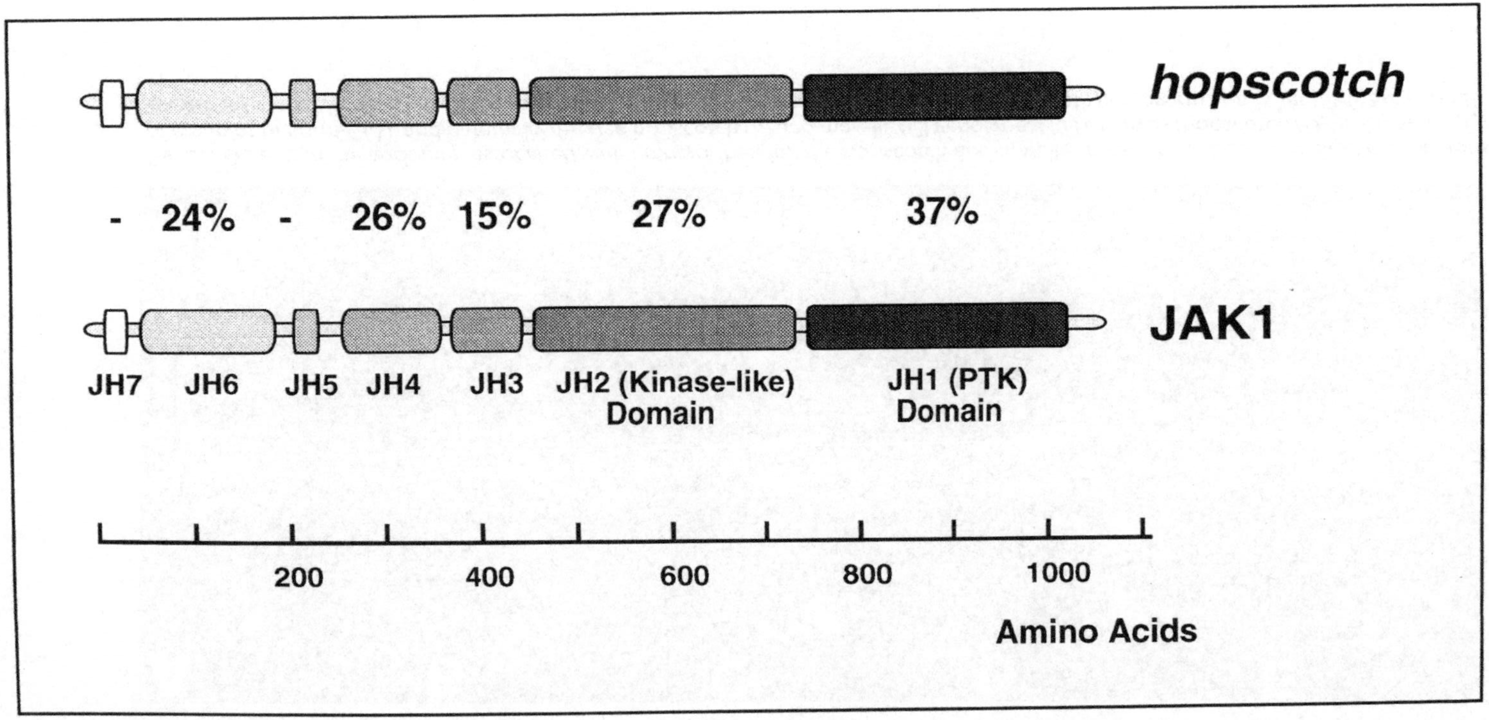

Fig. 25. Comparison of hopscotch *with mammalian JAK family members. Global comparison of the* hopscotch *protein with the mammalian JAKs. The domain structure is as shown for JAK1 in Figure 10. The figure represents the results of a sequence comparison between JAK1 and* hopscotch, *and underline the lower degree of similarity in regions outside the PTK domain (JH1).*

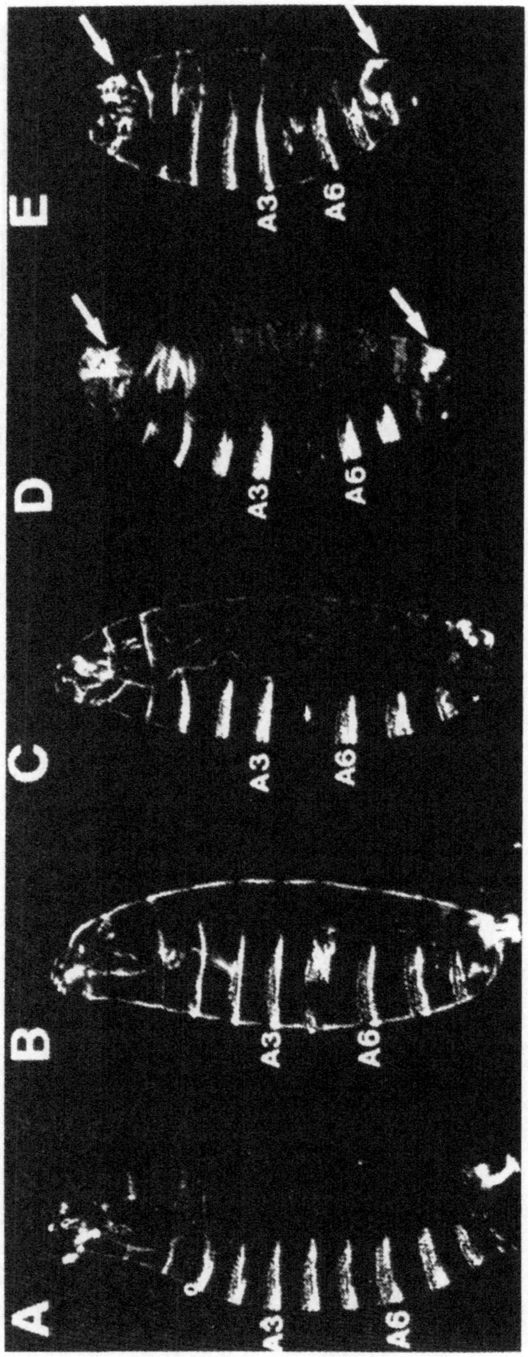

Fig. 26. Defects in the abdomen associated with embryos bearing the hopscotch and marelle mutations. A-E are dark field micrographs of embryos of normal (A), and paternally rescued hopscotch (B) and marelle (C) genotype. Null embryos (hopscotch (D) and marelle (E)) appear to have additional defects in the thoracic and head and tail regions (arrows). Reprinted with permission from: Cell 1996; 84:411-429.

HOPSCOTCH EMBRYOS EXHIBIT DEFECTS IN SEGMENTATION GENE EXPRESSION

The cuticular defects in *hopscotch* flies suggested that there may be a disturbance in the underlying mechanisms of embryonic segmentation. Thus, an examination of the expression patterns of segmentation genes such as the Gap genes and the Pair-rule genes was undertaken in both wild type and *hopscotch* animals, using both antibodies against these proteins, and the use of transgenic flies expressing *lac*Z downstream of an *even-skipped* promoter (see above). The expression patterns of the Gap genes *hunchback* (*hb*), *giant* (*gt*), *Krüppel* (*Kr*), and *knirps* (*kni*) in *hop* embryos are indistinguishable from those found in wild type flies. Thus *hop* is likely to act after the Gap genes exert their effect. Analysis of the expression patterns of three Pair-rule genes, namely, *runt* (*run*), *even-skipped* (*eve*) and *fushi tarazu* (*ftz*), showed interesting effects upon the expression of these genes in particular stripes within the embryo. Most notable was the complete loss of *run* expression in the 5th stripe of *hop* embryos, echoing the loss of the fifth abdominal (A5) segment which was the common feature of the phenotype of all *hop* mutants. Interestingly, the borders of the fourth and sixth stripes of *run* expression remain well defined.

The use of the *eve*5.2-*lac*Z construct as an indicator of *eve* regulation was discussed above (see Fig. 31). In wild type embryos, *lac*Z is expressed in stripes 2, 3 and 7. However, in *hop* embryos this *lac*Z transcription is almost abolished in stripe 3; although a low level of transcription in the ventral region is still observed. The use of the stripe 3-specific *eve* promoter element construct confirms the specific-stripe 3 loss in *hop* embryos. Thus, *hop* must exert its effect on *eve* stripe 3 via those transcription factors that bind the minimal element. The nature of the interaction is between Hop and the transcription factors which regulate Pair-rule gene expression is not known; however, it is tempting to speculate that it may be similar to the role that JAK family members play in regulating STAT-dependent promoters in mammalian cells.

In addition to effects on Pair-rule gene expression, *hop* embryos exhibit defects in the expression of the segment polarity genes *engrailed* (*en*) and wingless (*wg*), each of which are normally expressed in individual parasegments. In *hop* embryos the 10th stripe of *en* expression is almost completely missing, and the 9th stripe of *wg* expression is no longer present. Interestingly, these two stripes correspond to parasegments nine and ten which is precisely the region where defects in the fifth stripes of both *eve* and *ftz* occur, suggesting that the cuticle defects seen in this region of *hop* embryos are an accurate reflection of both *en* and *wg* misexpression.

Hopscotch is therefore a novel type of segmentation gene, and betrays the requirement of PTK activity for the transcriptional regulation of Pair-rule (such as *eve* and *ftz*) and segment polarity genes (such as *en* and *wg*). The ubiquitous expression pattern of *hop* transcripts contrast with the defined stripes of expression of the Pair-rule genes, however, and it seems likely that the regulation of Pair-rule gene expression by the Hop protein requires an additional activation step. At present, there are no candidate "cytokine" receptors which have been discovered in *D. melanogaster*. To affect that activation step, however, it may be that there are other mechanisms extant in the *Drosophila* embryo which might bring about that regulation.

TUMOROUS-LETHAL (*TUM-^{L}*)

The development of the *hop* system as a potential tool for a genetic analysis of this signal transduction pathway has been greatly facilitated by the discovery of a tumorigenic allele of this gene, named *Tumorous-lethal* (*Tum-^{l}*), which encodes a dominant upmutant of this PTK.[420] *Tum-^{l}* flies exhibit two phenotypes; formation of melanotic tumors and temperature-sensitive lethality. *Tum-^{l}* maps to a locus on the X chromosome precisely where the position of *hop* lies.[420] Deficiency mapping was used to define a lethality group, the members of which, except for *hop*, were rescued by *Tum-^{l}* revertant alleles. That *Tum-^{l}* is allelic to *hop* is recognized in the naming of this gene as *hop$^{tum\text{-}l}$*. In order to understand the intricacies of the *Tum-^{l}* phenotype, some discussion of the *Drosophila* immune system needs to be presented.

IMMUNITY IN *DROSOPHILA MELANOGASTER*

Drosophila melanogaster do not need an oxygen carrying blood cell, their oxygen requirements are fed by means of an intricate system of trachaea. Nonetheless, it has evolved a cellular immune system (based around a cell type known as hemocytes) to protect it from microbial infection.[421] During larval life, hemocytes are produced in a glandular, lobed structure at the anterior end of the heart-like dorsal vessel, a structure known as the lymph gland. Hemocyte lineage development is poorly understood, but 90% of larval hemocytes are spherical, phagocytic cells called plasmatocytes, capable of encapsulating foreign bodies and secreting anti-bacterial peptides and extracellular matrix. Upon pupariation, the plasmatocytes become competent to differentiate into lamellocytes, a flattened cell specialized in encapsulation. Melanotic masses, dark acellular inclusions, are formed when plasmatocytes and lamellocytes act in concert with crystal cells to encapsulate and melanize a foreign body. These melanotic masses are not detrimental to the animal and most adults carry a few.

THE *HOP^{TUM-L}* PHENOTYPE

hop^{Tum-l} mutants raised under standard culture conditions at 18°C are viable,[423] although there is over-proliferation of plasmatocytes, reaching levels 5-20 times wild type.[422] In contrast, at 29°C, hemi- or homo-zygous *hop^{Tum-l}* mutants die in late larval or early pupal life and contain large numbers of melanotic tumors.[422,423] The lymph glands increase in size and number, and premature differentiation of plasmatocytes into lamellocytes is seen. Hemocytes from animals raised at the restrictive temperature (29°C) possess altered surface properties and aggregate independent of foreign bodies into melanotic masses.[424] These cells invade tissues of the larvae, pupae and adult, causing degeneration. The source of this leukemia-like proliferation of hemocytes in *hop^{Tum-l}* larvae is the lymph gland.[422] Serial transfer experiments demonstrate that the organ bears a genuine neoplastic malignancy, causing host death even after seven serial transplantations. Interestingly, death appears to be independent of melanotic tumor formation; some subcultured *hop*$^{Tum\text{-}l}$ lymph glands lost the melanotic phenotype with passage, nonetheless, they remained lethal following transplantation.[422] Lymph glands taken from 29°C donors will kill their hosts even if the hosts were raised at 15°C, suggesting that an irreversible step has occurred in early larval life in the developing lymph gland due to *hop*$^{Tum\text{-}l}$ activity.[422]

The molecular basis for the *hop*$^{tum\text{-}l}$ phenotype

The molecular basis of the tumorigenic alteration of *hop* into *hop*$^{Tum\text{-}l}$ appears to be a single point mutation causing a glycine to glutamic acid change at amino acid position 341 (G341E) within the extra-catalytic domain of HOP.[419,425] Introduction of this mutation into a normal *hop* transgene caused the lethality and melanotic phenotype associated with *hop*$^{tum\text{-}l}$.[419] Analysis of HOP$^{Tum\text{-}l}$ in transfected *Drosophila* S2 cells demonstrates a several-fold increase in autophosphorylation over HOP, indicating an increased catalytic activity.[419] However, no change in activity was seen when culturing the cells at permissive (18°C) and non-permissive (25°C) temperatures, providing a molecular explanation for the persistent malignancy of transplanted lymph nodes in hosts raised at 15°C. Importantly, over-expression of normal *hop* throughout the developing larva caused tumor formation, hypertrophied lymph glands and death[419] suggesting that overactivity of HOP is responsible, and that *hop*$^{Tum\text{-}l}$ is a hypermorphic mutant, not a neomorphic one. The mechanism by which the G341E alteration causes the constitutively raised levels of PTK activity is at present unknown. Nonetheless, screens for extragenic second site suppressors of this locus facilitated by the easily scorable melanotic phenotype will no doubt be a fertile source of other molecular components of this pathway. The melanotic phenotype of the *hop*$^{Tum\text{-}l}$, although not responsible for death of the fly per se, is a useful marker to enable screens for suppressors or enhancers of the *hop*$^{Tum\text{-}l}$ activity. Several groups have pursued this line of inquiry, claiming to have isolated a dozen suppressors and one enhancer.[420,425]

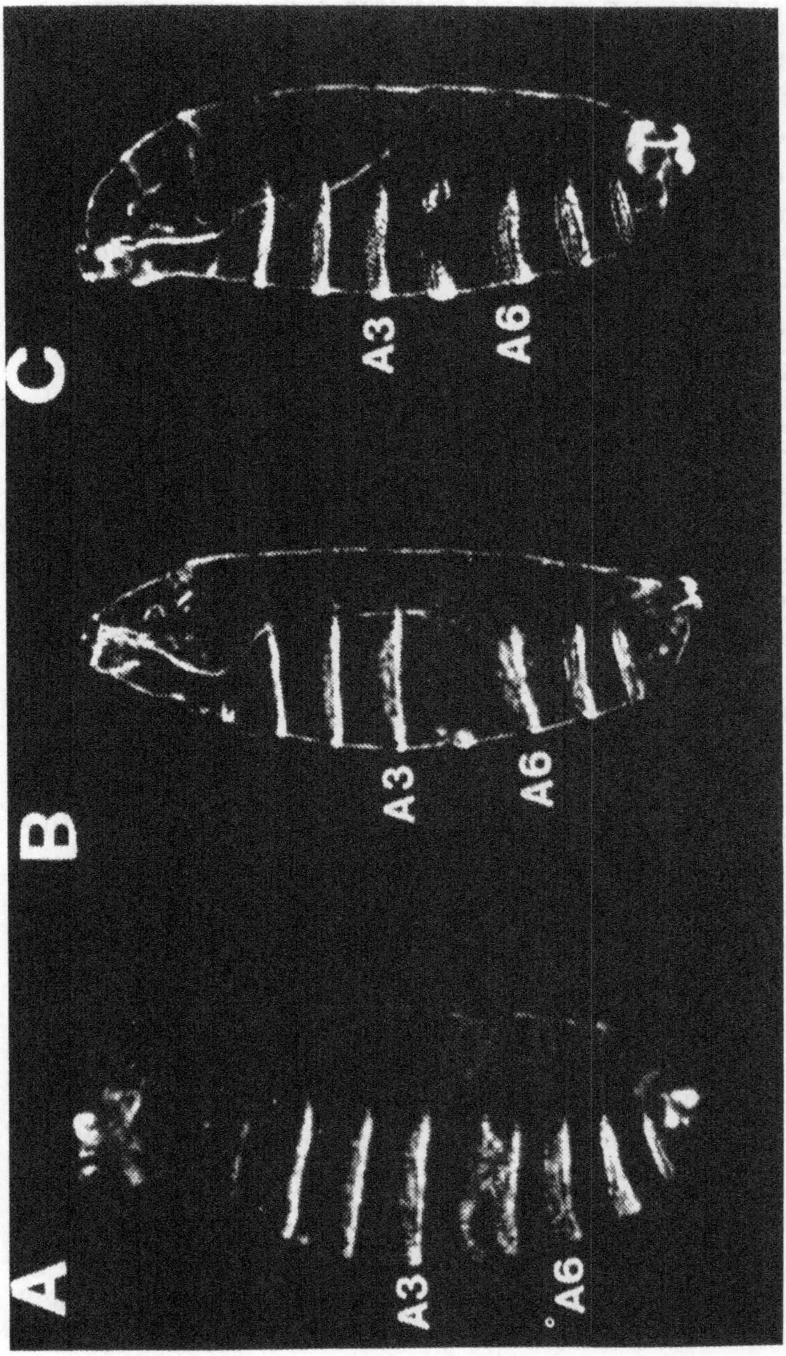

Fig. 27. Enhancement of hopscotch phenotype by marelle. Dark field cuticle pattern of embryos derived from females with defective allele of hopscotch. (A) hopscotch embryo, (B & C) hopscotch embryos carrying a heterozygous marelle mutation. Note the loss of segments A5 and A4 in the double mutant embryos. Reprinted with permission from: Cell 1996; 84:411-429.

DISCOVERY OF A *DROSOPHILA* STAT PROTEIN

The *Drosophila* STAT protein was uncovered in two independent genetic screens of the *Drosophila* genome. In the first, a screen of maternal effect phenotypes which were similar to the originally described *hopscotch* phenotypes (namely that abdominal segments five and parts of abdominal segment four were missing), was undertaken by the same laboratory that discovered the *hopscotch* gene.[426] Thus, in a large screen for zygotic lethal mutations, a P-element induced mutation known as L(3)06346 was discovered and was mapped to 92E2-4. The maternal effect segmentation phenotype of this mutant was very similar to the effect of loss of the *Hopscotch* gene activity during embryogenesis, the mutant was given the name *marelle* (which is the French word for "hopscotch"). The notion emerged that this might be a gene in the same pathway as the *hopscotch* gene previously described. *Marelle (mrl)* was shown to be required for zygotic viability since homozygous null flies died as larvae (Fig. 27). The dead larvae had a normal cuticle pattern, but possessed imaginal discs that were reduced in size, suggesting a zygotic role for *marelle* in cellular proliferation. In embryos derived from females lacking *mrl*, a defective segmentation pattern was observed which closely resembled the *hopscotch* phenotype. It was further observed that *mrl* was epistatic to *hopscotch*. Hence the absence of an *mrl* gene resulted in the inability of the dominant temperature sensitive *hopscotch* allele ($hop^{Tum\text{-}l}$), to produce melanotic tumors in emerging adults. These data strongly supported the notion that *mrl* encoded a component of the *hopscotch* signal transduction pathway and the *mrl* protein functioned downstream of the *HOP* protein tyrosine kinase.

The molecular cloning of the *mrl* gene was accomplished by means of the plasmid rescue of P-element insertion associated clones. The gene located at the site of the P-element insertion, was shown to be encoded by an abundant RNA species of 4 kb and by two less abundant species of 3.5 and 3.0 kb, respectively. The RNA encoded by the *mrl* gene was expressed throughout embryonic, larval, pupal and adult stages of the *Drosophila* life cycle. The two smaller RNAs were detected during larval stages only. In situ hybridization studies using a cDNA encoding *mrl* showed that there was a uniformly high level of *mrl* expression during early syncytial and cellular blastoderm embryos—a pattern remarkably similar to that exhibited by *hopscotch.* During germ-band extension, *mrl* expression was detected in a stripe pattern within every segment of the developing embryo.

The protein encoded by the *mrl* gene was demonstrated to possess 37% identity with the cDNA for human STAT5.[426] The overall similarity between these proteins is remarkably high and extends to all domains identified and described in chapter 4, on the STAT proteins (see Fig. 28).

A second group contemporaneously isolated a *Drosophila* STAT cDNA by application of the polymerase chain reaction using degenerate oligonucleotides from highly conserved elements held in common between all the mammalian STATs.[427] The identity of the putative coding sequence of the *Drosophila* protein thus discovered, was shown to be about 35% identical to STAT5 and STAT6, and exhibited between 25-30% identity to other STAT family members. The highest regions of identity were shown to be in the SH2 domain (Fig. 30). Interestingly, this is the region in which the greatest similarity also occurs between the mammalian STAT proteins. This second publication also showed very similar data concerning the chromosomal mapping of D-STAT and the expression pattern of D-STAT from cellular blastoderm through to germ-band extended embryos. It was subsequently shown in this study that D-STAT was allelic to a suppressor of the tumorous lethal version of the *hopscotch* protein suggesting that it was in the same pathway as the *HOP* protein (Figs. 29, 30).

A D-STAT DNA binding site was uncovered by the process of optimum binding site selection and was shown to be encoded by the following sequence:

TTT CCC GG AAA

Finally, the requirement for phosphorylation at tyrosine 704, was shown for DNA binding of the *Drosophila* STAT protein.

Both studies[426,427] then undertook to investigate the expression of Pair-rule genes in both normal and STAT mutant embryos (Fig. 31). In these studies a particularly interesting effect on the Pair-rule *even-skipped (eve)* was uncovered, thus the abolition of stripe 3 expression of the *even-skipped* gene demonstrated that both

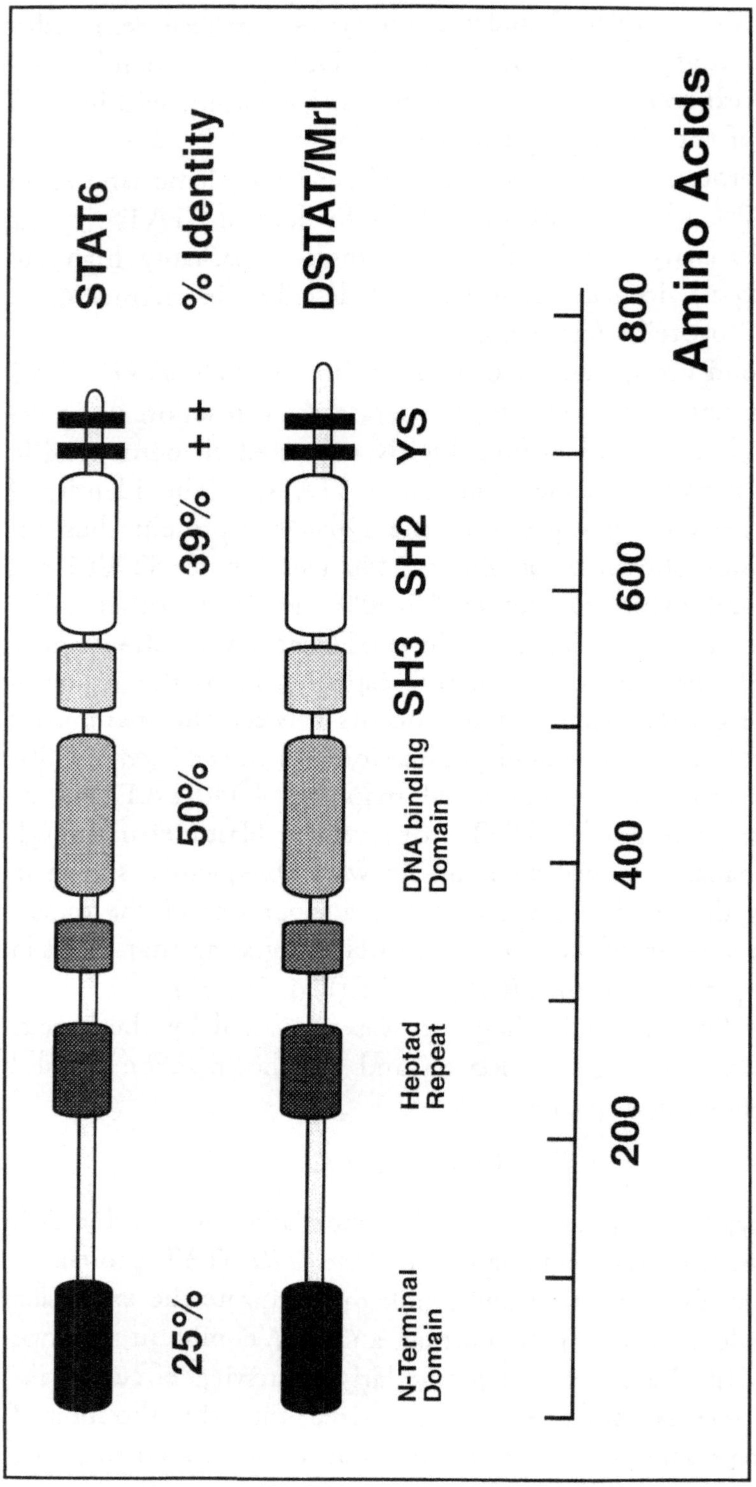

Fig. 28. Comparison of marelle with mammalian STAT family members. Global comparison of the marelle protein with the mammalian STATs. The domain structure is as shown for STATs in Figure 13. The figure represents the results of a sequence comparison between STAT6 and marelle. Note the high degree of similarity in the putative DNA binding domain.

```
STAT1    TNGSLAAELRHLQLKEQK-NAGNRTN-EGPLIVTEELHSLSFETQLCQPG--LVIDLETT    451
STAT6    PGNCCSALFKNLLLKKIK-----RCERKGTESVTEEKCAVLFSASFTLGPGKLPIQLQAL    406
DSTAT    NNHVFSASFRNMQLKKIK-----RAEKKGTESVMDEKFALFFYTTTTV--NDFQIRVWTL    476

STAT1    SLPVVVISNVSQLPSGWASILWYNMLVTEPRNLSFFLNPPCAWWSQLSEVLSWQFSSVTK    511
STAT6    SLPLVVIVHGNQDNNAKATILWDNAFSEMDRVP--FVVAERVPWEKMCETLNLKFMAEVG    463
DSTAT    SLPVVVIVHGNQEPQSWATITWDNAFAEIVRDP--FMITDRVTWAQLSVALNIKFGSCTG    534
```

Fig. 29. Comparison of STAT1, STAT6 and marelle *(D-STAT) in the region of the DNA binding domain. Amino acid positions of the most C-terminal residue are shown on the left of each line.*

```
STAT1    DTILELIKNDLLCLWNDGCIMGFISKERERALLK--DQQPGTFLLRFSESSREGAITFTWVE    618
STAT6    DGVLDLTKRCLRSYWSDRLIIGFISKQYVTSLLL--NEPDGTFLLRFSDSE-IGGITIAHVI    577
DSTAT    FAIMKLTKDHMLGMWKAGCIMGFINKTKAQTDLLRSVYGIGTFLLRFSDSE-LGGVTIAYVN    640

                  --βA--       --αA--       --βB(FLVRES Motif)--

STAT1    RSQNGGEPDFHAVEPYTKKELSAVTFPDIIRNYKVMAAENIPENPLKYLYPNIDKDHAFGKYYS  682
STAT6    RGQD-GSPQIENIQPFSAKDLSIRSLGDRIRDL----------AQLKNLYPKKPKDEAFRSHYK  630
DSTAT    -----ENGLVTMLAPWTARDFQVLNLADRIRDLDVLCW---LHPSDRNASP-VKRDVAFGEFYS  695
```

Fig. 30. Comparison of STAT1, STAT6 and marelle *(D-STAT) in the region of the SH2 domains. The location of two of the β-sheets and one of the α-helices of a typical SH2 domain are shown. Amino acid positions of the most C-terminal residue are shown on the left of each line.*

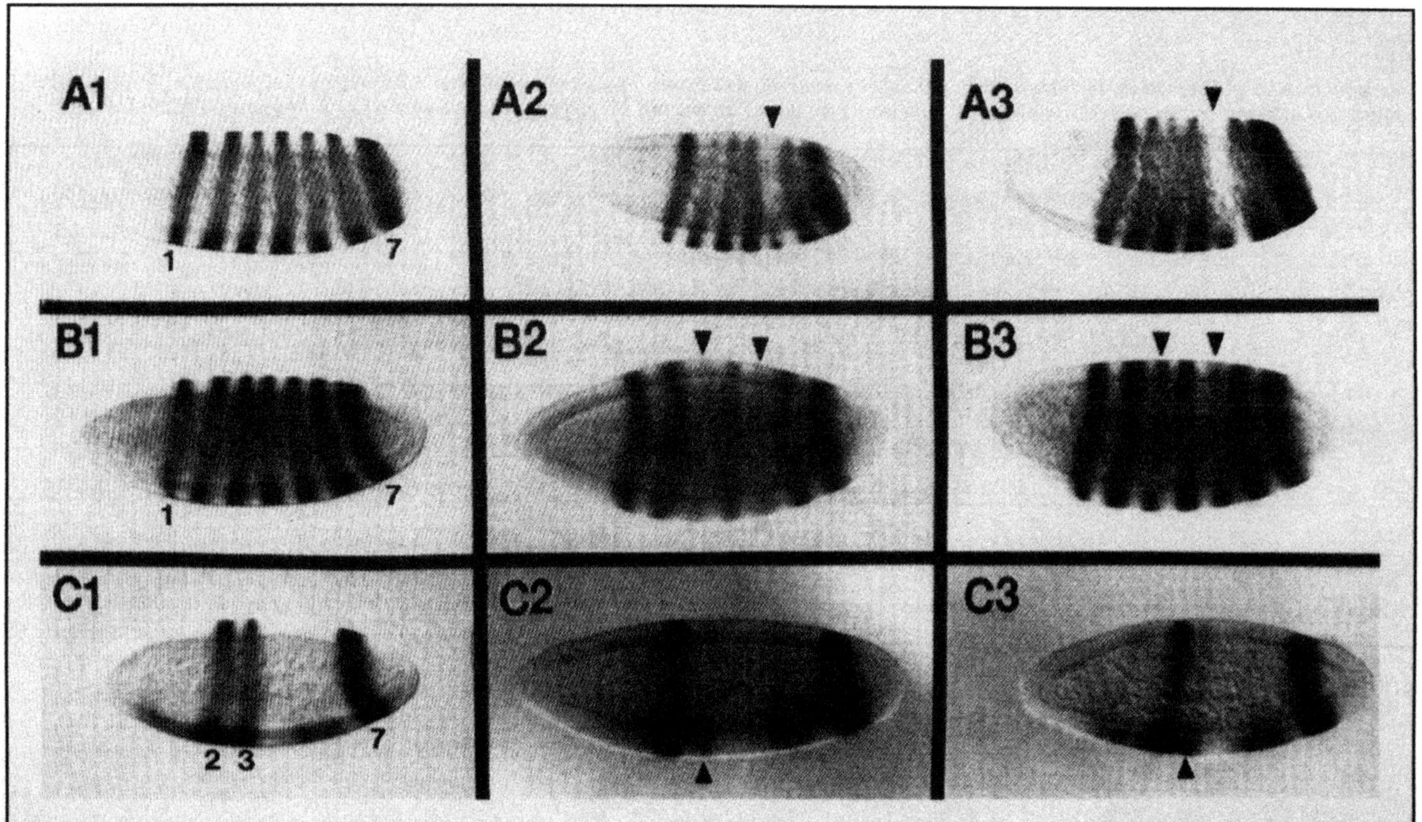

Fig. 31. Pair-rule gene expression in wild type, hopscotch *and* marelle *Drosophila embryos. The expression pattern of the pair-rule gene* runt *is shown, as revealed by in situ hybridization. The embryo shown in panel A1 is wild type, whereas those shown in A2 and A3 are mutant* hopscotch *and* marelle *embryos. Expression of* even-skipped *transcripts by embryos of wild type (B1),* hopscotch *(B2) and* marelle *(B3) genotype are also shown. Panel C (C1, C2 and C3), shows the expression of the reporter gene* lacZ*, driven by an enhancer from the* even-skipped *promoter. Note that the expression of* lacZ *is missing from stripe three in both* hopscotch *and* marelle *embryos. Reprinted with permission from: Cell 1996; 84:411-429.*

hopscotch and *even-skipped* and *D-STAT/mrl* were in the same pathway which controlled the stripe 3 specific expression of the *even-skipped* gene, subsequently showing that there are two consensus STAT binding elements in the *even-skipped* promoter. Single mutations in these STAT binding elements significantly reduced expression; however, the mutation of both elements resulted in the complete abolition of stripe 3 and also stripe 7 expression of the *even-skipped* gene. The result of the observation that stripe 7 expression was reduced was not expected since the formation of *even-skipped* stripe 7 was not significantly affected in the *hopscotch* mutant embryos.

Two possible models can now be built concerning the extent of the JAK-STAT pathway in *Drosophila melanogaster*. Firstly, the stripe specific expression of *even-skipped* suggests that there is a stripe specific manifestation of the activated JAK-STAT pathway during embryogenesis in this region. This could result from one of two distinct mechanisms: either a receptor ligand system, perhaps similar to the cytokine receptor system present in the mammalian cells, or by some other as yet unidentified mechanism, might result in this stripe specific activation of the JAK-STAT pathway in *even-skipped* stripe 3. Conversely, the signal need not be spatially restricted and might extend throughout the entire embryo, but rather there might be a decreased level of factors which repress the stripe 3 enhancer of *even-skipped* only at the level of stripe 3, which in turn might result in the stripe 3 specific expression of the *even-skipped* gene.

EXTENT OF OUR UNDERSTANDING OF THE JAK-STAT PATHWAY IN *DROSOPHILA MELANOGASTER*

The power of *D. melanogaster* as a genetically tractable organism has already been exploited to help fit the JAK-STAT pathway into the overall signaling network of the cell. $hop^{Tum\text{-}l}$ produces a protein which delivers continual stimulus for differentiation and proliferation, independent of extracellular signals. This dominant effect is suppressed in animals carrying the *prune* (*pn*) or *altered wing disc/killer of prune* (awd^{Kpn}) mutations.[428] Interestingly, *pn* was originally discovered as an eye color mutant. The presence of this allele results in a decrease of larval pteridenes,[435] the biosynthetic precursor of which is GTP. *pn* is thought to encode a GTPase activating protein (GAP)-like molecule.[429-431] *awd*^{Kpn} is a neomorphic mutation of the *Drosophila melanogaster* nm23 homologue *awd.* This gene encodes a nucleotide diphosphate kinase responsible for conversion of GDP to GTP.[432-434] *pn* exhibits a lethal interaction with *awd*^{Kpn} which has similarities with the human genetic disease neurofibromatosis (NF1).[435] The tumor suppressive effects of *awd*^{Kpn} are specific for the *hop*$^{Tum\text{-}l}$ allele[428] while other melanotic mutants are not suppressed. On genetic grounds, this interaction strongly implies a direct crosstalk between G-protein signaling and the JAK-STAT pathway.

In summary, the way is now open for the introduction of the sophisticated genetic approaches available in genetically tractable organisms such as *melanogaster*, for the analysis of the JAK-STAT pathway. Epistatic analyses of existing mutant strains with *hop* or *hop*$^{tum\text{-}l}$ are already underway, and second site mutagenesis protocols based around the phenotype of the dominant *hop* allele, *hop*$^{tum\text{-}l}$ are close at hand. The prospect of placing the JAK-STAT pathway into the signaling network of eukaryotic cells is indeed an exciting one.

CHAPTER 9

Future Trends

In the preceding chapters we focused upon the essential elements of the JAK-STAT pathway; the molecules themselves; the order of events by which these molecules are recruited for the transmission of cytokine dependent signals; and the structural modules (SH2 domains, PTK domains etc.) that coordinate the protein-protein interactions which underpin these signal transduction processes. While a number of gaps in this story need to be filled in, the most pressing questions relate to the integration of the JAK-STAT pathway into the network of pathways by which cells coordinate their physiological responses to extracellular stimuli. In this final chapter, we will attempt to show how the JAK-STAT pathway is likely to become integrated into the overall fabric of the cellular signal transduction network.

INTEGRATING THE JAK-STAT PATHWAY INTO THE CELLULAR SIGNALING NETWORK

In the opening chapter of this book we discussed the *ras*-MAP kinase pathway. Fifteen years of progress in the area of oncogene research, coupled with a concerted effort aimed at understanding the biochemistry of the proteins which contribute to this pathway, have combined to produce what appears to be a central pathway of signal transmission within eukaryotic cells. Intuitively, one would expect that this pathway should interact in some way with the other major pathways within the cell, otherwise true integration of signal transduction could not occur. The recent melding of the JAK-STAT pathway with the *ras*-MAPK pathway is thus an experimental confirmation of an anticipated networking of the intracellular circuitry of the cell.

Common Components of Cytokine Receptor and RTK Signaling

We have already discussed at length the notion of cytoplasmic PTK recruitment by activated cytokine receptors. The central role that these activated PTKs play in the triggering of particular signal transduction pathways has provided a basis for the notion that the mechanics of signal transduction downstream of both cytokine receptors and RTKs is thematically very similar. Essentially, both are based around the recruitment of SH2-containing signal transduction molecules. It is clear now that not only are there mechanistic similarities, but there is functional overlap in the components recruited by each of these pathways. For example, the binding of ligand to the PTK receptors for CSF-1,[388] PDGF[388] and EGF[250] results in the tyrosine phosphorylation of particular members of the JAK family, and, not surprisingly, ultimately activates STAT-containing transcription factor complexes. Conversely, overlap further extends to the downstream signaling molecules employed by the cytokine receptors, where in fact, a number of intracellular signaling molecules thought previously to be activated exclusively by the RTKs, have also recently been identified in association with cytokine receptors. EPO, IL-3 and steel factor stimulate SHC activation,[108,436] EPO itself activates Raf-1,[437] p21*ras*,[162] GAP[162] and PI3'-K,[438-440] while the IL-4 receptor has been shown to associate with the nonreceptor PTK, *fes*.[441]

Cross-Talk Between the *RAS*-MAP Kinase Pathway and the JAK-STAT Pathway

The possibility of an interaction between the ras-MAPK pathway and the JAK-STAT pathway has been a question for some time. Until recently, only the role of tyrosine phosphorylation in the activation of the STAT complexes had been examined, but more recently, the evidence for the requirement for serine phosphorylation in this same activation has become available. The IFNγ-induced differentiation of promonocytes into mature macrophages involves the serine phosphorylation of STAT1α containing GAF, without any obvious increase in tyrosine phosphorylation of this STAT.[249] This additional modification in conjunction with tyrosine phosphorylation may in some way explain the apparent increase in DNA-binding activity of GAF following monocyte differentiation when compared to the same measurements in extracts from an undifferentiated cell type. The activation of STAT1-containing transcription complexes in response to IFNγ and of STAT3-containing transcription complexes in response to EGF[241] or IL-6 and CNTF,[254] also appears to require serine phosphorylation for maximum efficiency. These studies together predicted[254] and confirmed the presence of a serine residue embedded in a candidate MAPK phosphorylation site[241] in STAT1, and a similarly critical serine residue in STAT3.[241] While this observation was tantalizing enough to suggest the physiological involvement of a MAPK in this activation, it still remained to be demonstrated that these serine residues were bona fide MAPK targets in vivo, and that MAPK was directly linked to the JAK-STAT pathway and hence, STAT activation.

A direct biochemical link between the activation of MAPK, specifically the 42 kDa form, known also as extracellular signal-regulated kinase 2 (ERK2) and IFNβ-stimulated activation of STAT proteins[442] emerged in recent publications from a number of laboratories. In these studies, fusion protein association experiments demonstrated that this isoform of MAPK interacts constitutively with the 50 amino acid membrane proximal region of the α subunit of the IFNα/β receptor. Following IFNβ stimulation, there also appears to be a time-dependent association with STAT1α.[442] The question raised by these findings was whether this apparent interaction was coincidental, or did these two proteins impinge upon the same pathway? This question was addressed when a nonactivatable MAPK protein (generated by mutation of the activating phosphorylation sites) was shown to inhibit IFNβ-stimulated gene activation from an ISRE DNA element, confirming that the p42 MAPK was actually signaling upstream of STAT1α in the IFNβ pathway.[442] Thus, it appears that the observation that signaling downstream of the IFNγ receptor requires the involvement of serine phosphorylation (in conjunction with the obligatory tyrosine phosphorylation) for maximal STAT activation,[241] is not an epiphenomenon, but rather a pivotal component of the activation process.

It is becoming evident that the convergence of both the RTK-activated and the noncatalytic cytokine receptor-activated pathways are utilizing common evolutionarily successful themes and components in order to fine tune the specificity of their responses (Fig. 33). We can now no longer presume that there is mutual exclusivity of these pathways. The interaction of the JAK-STAT pathway and the ras-MAPK pathways is shown in Figure 32.

THE JAK-STAT PATHWAY IN PATHOLOGIES

JAK-STAT PATHWAY AND ONCOGENIC TRANSFORMATION BY ACTIVATED PTKs

The oncogenic activation of cytoplasmic and transmembrane PTKs, by means of capture by retroviruses, or by means of the acquisition of other activating mutations, has provided a useful platform from which to investigate the signaling pathways in which these proteins are involved. The frequency with which members of the tyrosine kinase family are involved in such oncogenic processes serves to highlight the importance of tyrosine phosphorylation (and its regulation) in the coordination of essential cellular functions. Oncogenic activation of tyrosine kinases typically involves the constitutive activation of normally tightly controlled catalytic activity. Cellular transformation by retrovirally captured oncogenes is a result of the constitutive activation of a number of intracellular signal transduction pathways, and thus it is perhaps not surprising that the transformation by v-SRC[443] and v-ABL[444] activates the JAK-STAT pathway.

v-SRC is the avian retroviral version of the cytoplasmic nonreceptor PTK *c-SRC*.[445] The net result of retroviral capture has been the deletion of the C-terminal regulatory tail of the *c-SRC* gene, resulting in a constitutively activated tyrosine kinase domain.[445] Mechanistically, the C-terminal tail of the SRC family of PTKs encompasses a tyrosine residue (tyrosine 527) required to be phosphorylated to negatively regulate the activity of the PTK,[446] by interaction with the SH2 domain of a second SRC molecule (by extrapolation from what has been demonstrated for the *lck* PTK[28,447]). Recent studies on fibroblasts expressing the Rous sarcoma virus form of *v-SRC* exhibited a constitutive DNA-binding

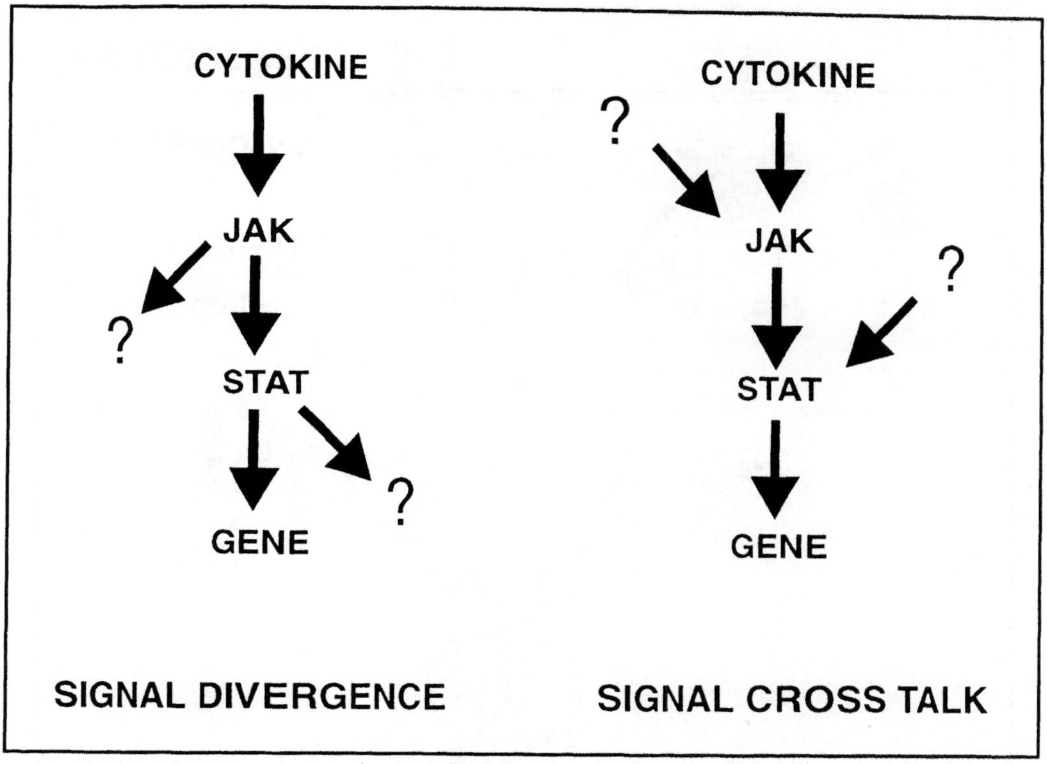

Fig. 32. Burning questions about the JAK-STAT pathway. Are there other outputs from the components of the JAK-STAT pathway that do not involve their partners? Is there signal divergence? Similarly, are there other signaling pathways feeding into the JAK-STAT pathway? Is there signal cross talk? (diagram based upon one devised by A.Oates)

activity displaying an affinity for an SIE element and the APRF response element (APRE).[444] The *v-SRC*-induced SIE-binding factor comigrated with the SIF-A factor, and indeed this novel activity contained tyrosine phosphorylated APRF, or STAT3.[444] This study addressed relative potency of various forms of activated SRC, and it appeared that the amount of constitutive binding activity, and the level of tyrosine phosphorylation, increased with increasing SRC potency.[444]

The Abelson murine leukemia virus (A-MuLV) oncogene *v-ABL* is also a PTK. In murine pre-B lymphocytes, *v-ABL* not only induces constitutive activation of the STAT proteins, but also tyrosine phosphorylation of JAK1 and JAK3.[448] The A-MuLV form

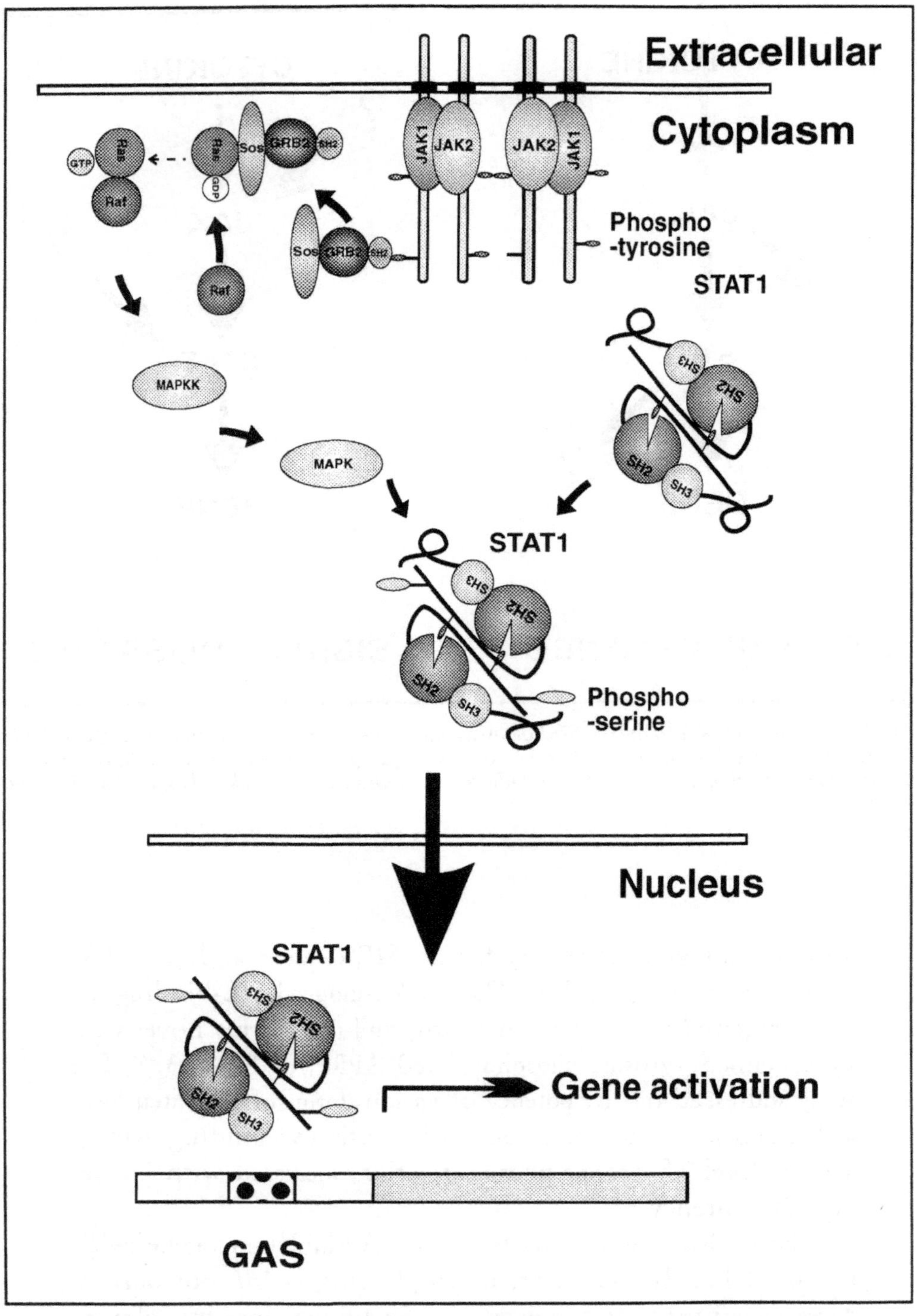

Fig. 33. Integration of the JAK-STAT pathway with the ras/MAPK pathway. One possible scenario whereby these two pathways may converge is shown. Details are discussed in the text.

of *v-ABL* exists as a fusion between the viral Gag gene and the C-terminus of the nuclear-localized form, *c-ABL*, which includes the PTK and SH2 domains, but not the SH3 domain.[449] Interestingly, *c-ABL* appears to be normally resident in the nucleus, but the loss of the SH3 domain contributes to the relocalization of *v-ABL* to the cytoplasmic compartment, an event implicated in the transforming capabilities of this protein. *v-ABL* is also responsible for the induction of cytokine-independent growth in hemopoietic cells (without an apparent autocrine mechanism).[450]

It is well established that IL-4 and IL-7 signal transduction requires the activation of JAK1 and JAK3,[358] and as pre-B cells require the addition of IL-7 for growth maintenance,[451] it is not surprising that growth-factor independence induced by *v-ABL* transformation of these cells results in the constitutive activation of the components of the IL-7 signal transduction pathway. Such transformation activates a DNA binding complex capable of binding a GAS element, an activity reminiscent of those typically seen following IL-4 and IL-7 stimulation.[448] These complexes contain STAT5 and STAT6. Not only are there activated STAT complexes in these transformed lymphocytes, but JAK1 and JAK3 also exhibit evidence of increased activity.[448] Furthermore, coimmunoprecipitation experiments suggest a direct interaction between *v*-ABL and both activated JAKs,[448] raising the possibility of complex formation between members of these two classes of PTK.

In summary, it is evident that the unbridled kinase activity of a transforming PTK can have profound effects upon many of the intracellular signal transduction pathways within a cell. Clearly, both *v-SRC* and *v-ABL* are capable of recruiting members of the STAT family into functional DNA binding complexes, and this may well be part of the oncogene's mechanism of transformation, nonetheless, this needs to be formally proven in order establish that this is more than an epiphenomenon associated with a hyperactive PTK.

HTLV-I and the Activation of the JAK-STAT Pathway

The retroviral immortalization of human $CD4^+$ T cells by human T cell leukemia virus type I (HTLV-I) leads to a shift from IL-2-dependent to IL-2-independent growth. As described elsewhere, the JAK-STAT pathway has been firmly implicated in IL-2 signaling, and has been demonstrated to be absolutely required for the transmission of this signal. It appears that the HTLV-I-induced IL-2 independence involves the constitutive activation of the JAK-STAT pathway.[443] In this study, the loss of IL-2 responsiveness subsequent to immortalization by HTLV-I suggests the loss of the requirement of receptor activation for cell survival (confirmed by the lack of IL-2 production by these cells), and that the activation of the intracellular signaling pathway downstream of the IL-2 receptor (or different receptors in other cases) is sufficient to maintain cell survival.

HTLV-I transformation of peripheral blood lymphocytes (PBLs) leads to the constitutive (and IL-2-inducible for a period of time prior to full IL-2-independence) activation of a transcription factor complex that recognizes a GAS DNA element from the FcγRI promoter.[443] STAT3 and STAT5 are activated in response to IL-2 in PBLs,[146] and are present in these activated GAS-binding complexes arising from the HTLV-I transformation.[443] In agreement with the observation that the IL-2 pathway becomes activated, both JAK1 and JAK3 are constitutively phosphorylated on tyrosine in these cells.[443] During the initial phase of residual IL-2 responsiveness, this tyrosine phosphorylation of STAT3 and STAT5 and JAK1 and JAK3 was augmented by IL-2 stimulation. Unlike *v-ABL* and *v-SRC*, HTLV-I does not possess any intrinsic PTK activity[452,453] that may be the active component of this pathway, thus an alternate mode of activation must be employed by this retrovirus to render the JAK-STAT pathway active.

Severe Combined Immunodeficiency (SCID)

A role for JAK3 in mitogenesis has been demonstrated via the IL-2R chain.[175] It is clear that biological activity of the IL-2 receptor absolutely requires coupled tyrosine kinase activity,[454] and the membrane-proximal regions of the receptor (including Box 1) are associated with PI3'-K activation and the activation of a tyrosine kinase.[454] One of these PTK activities has been defined as JAK3.[175] The biological role of IL-2 signaling has been further elucidated with the identification of JAK1 and JAK3 involvement in this as well as other IL-2 receptor γ_c chain-utilizing receptor complexes pathway.

It has recently come to light that there is an important link between JAK3 and the development of a disease condition—X-linked severe combined immunodeficiency (XSCID). Defects in the common γ_c receptor chain essential for signaling by IL-2,[134] IL-4,[140,149,455] IL-7,[141,142] IL-9[149,455] and IL-15,[143] lead to the development of XSCID, a disease defined by impaired function of B cells, and a T cell deficiency.[148,456] XSCID in one family appears to result from the disruption of IL-2 and IL-4-mediated responses, and has been shown to be directly linked to mutations in the common chain for each of these two receptors, namely IL-2Rγc. Affected members of the aforementioned family have a single nucleotide substitution (Leu 271 to Gln) within the cytoplasmic domain of the common chain, and more specifically, within the proposed SH2-homology subdomain.[149,455] This point mutation led to a severe decrease in the association of JAK3 with this receptor chain, suggesting a local structural requirement of this amino acid for an interaction with JAK3.[149,455] While PTK activity of JAK3 in this situation was not investigated, it is reasonable to assume that it would be reduced concomitantly with diminished receptor association, thereby correlating association, and functionality with signaling. Indeed, fibroblasts which do not express JAK3, when reconstituted with IL-2R failed to signal via IL-2 until JAK3 was introduced into the cells, further confirming a requirement for JAK3 in IL-2-mediated signal transduction.[107] A decrease, but not total abrogation, of the association of JAK3 with this mutated chain also suggested that other regions of the receptor chain may be required for JAK3 association, perhaps with lower affinity than that seen around the SH2-homology subdomain. This has been demonstrated by means of generating gross deletions of the intracellular domain of the γ_c chain. Removal of all but four membrane-proximal amino acids, or all but the SH2-homology subdomain of the intracellular domain, abrogated JAK3 association. Thus, the SH2-homology subdomain is not the only element required for JAK3 association, but cooperative association sites are present in the C-terminus of the receptor.[107] The issues of structural or functional disruption of the receptor within these deletion mutants, as discussed for the GH receptor, must still be addressed for the IL-2R chain, and the true nature of the absolute requirement of the JAK kinases in this pathway remains to be elucidated.

Further mutations implicating JAK3 in SCID have recently been identified within the JAK3 gene itself. These mutations were shown to be distinct from each other, and resulted in different JAK3 products. A female patient with SCID had normal γ_c expression but did not express JAK3 protein. Inspection of both alleles of JAK3 in this patient revealed two different point mutations (inherited from each parent), each resulting in premature termination codons. The first mutant allele possessed a single base insertion in JH4[166] leading to a frame shift and premature stop codon, while the other was a nonsense mutation in JH2.[166,455] The inability of IL-4 to activate STAT6 in cells derived from this patient suggested a mechanism underlying her condition, and (incidentally) further demonstrated the link between STAT6 and JAK3 in the IL-4 signaling pathway.[455] JAK3 appears only to be associated with the IL-2 receptor γ_c chain and has yet to be implicated with any other receptor chain. This observation leads one to conclude that the phenotypic effects of the SCID (or XSCID) condition, whether it involves a receptor chain defect or not, is in one way or another, a consequence of lack of JAK3 function.

Two unrelated T cell negative, B cell positive SCID patients were also found to have JAK3 mutations leading to reduced JAK3 levels.[457] Both unrelated patients had consanguineous parents, and each patient displayed a different mutation. The first of these being a single point mutation of tyrosine 100 to cysteine in a JH7 conserved tyrosine, while the second was a deletion of 151 bp from the JH2 domain leading to a premature stop codon.[457] Interestingly, these mutations lead to an effect on T cell but not B cell development, and a reduction in NK cells.

The analysis a number of SCID and XSCID patients with mutations within the common γ_c chain or JAK3, have allowed the dissection of the function of both of these components in lymphoid development. Patients with SCID and JAK3 deficiency show a normal or increased level of B cells, whereas those with γ_c defects are basically devoid of B cells.[458,459] It is clear that T cell and NK cell numbers are diminished with a JAK3 defect, thus linking the requirement of JAK3 with T cell development.[455,457]

From studies such as these, it is apparent that the restricted expression pattern, and selective receptor-association of JAK3 allows the assignment of a role for JAK3 in lymphoid development, and the likely implication of JAK3 defects in the XSCID or SCID syndromes. It still remains, however, that the absolute contribution of JAK3 PTK activity and the requirement of the IL-2 receptor γ_c chain in such lymphoid development need to be clearly defined.

INTERACTION OF STATs WITH OTHER TRANSCRIPTION FACTORS

The recent detection of an interaction between the JUN transcription factor and a truncated form of STAT3 (STAT3β)[223] heralded a new chapter in the study of these proteins. JUN belongs to the family of basic-leucine zipper proteins and is capable of forming homodimers, or forming heterodimers with members of the Fos family of transcription factors.[460] In recent studies, by dint of the use of the yeast two-hybrid screening system,[461,462] it was demonstrated that the N-terminal portion of the c-JUN transcription factor (a region within which reside a number of domains known to be important for gene activation[463,464]) interacted with a truncated form of STAT3, namely STAT3β.[223] This complex was able to induce gene activation through the promoter of the α_2-macroglobulin gene (comprising an IL-6 response element [a STAT binding site] and a putative AP-1 [a JUN binding element] site).[223] Thus, a functional transcription factor complex composed of two mutually exclusive classes of transcription factor appear to cooperate in the activation of immediate early genes.

cDNAs encoding STAT3β have been isolated, and examination of the sequence revealed that this protein was identical to STAT3α, except that 55 amino acids were missing from the C-terminus of the long form of STAT3,[204,206] with a seven amino acid alternative C-terminus replacing the usual C-terminus.[223] Transfection experiments demonstrated that STAT3β/JUN cooperativity required both a functional IL-6 response element and an AP-1 site,[223] clearly confirming the reality of this interaction.

The notion of the DNA element "half-site" has been discussed in chapter 5, and the idea that composite elements made up of half-sites with different specificities is one that is particularly pertinent in light of the observed association of STAT3β/JUN. If the type of heteromeric transcription factor complex seen in ISGF3α and the STAT3β/JUN complex extends to the involvement of other STATs in other transcription factor complexes, then the complexity of response that can be elicited by a given cytokine is indeed immense.

NEW DATA EMERGING FROM STAT AND JAK KNOCK-OUTS

The advent of gene knock-out technology using mouse embryonic stem cells[465] has brought a harvest of important new data relating to the in vivo function of mammalian genes. Components of the JAK-STAT pathway are now being added to the list of genes knocked out, and the data emerging are providing unexpected insights into the role(s) of these molecules in normal development and disease.

Mice lacking JAK3 have been generated by two different laboratories.[466,467] Several studies highlight the fact that there is still some ambiguity regarding the role of JAK3 and the γ_c chain in B cell development. In one study the first coding exon of JAK3 was targeted,[467] whereas in another study[466] the portion of the gene encoding the region of JAK3 lying between the PTK domain and the kinase related domain was targeted. The phenotype of the mice in both studies was essentially identical. In these mice, the absence of JAK3 had no impact upon their overall growth and development. However, profound alterations in T and B cell development were observed. Thus, the JAK3 –/– mice exhibited a lack of immature and mature B cells, reflecting a block in the development of B cells at the pre-B stage in the bone marrow.[466] Similarly, the thymuses of the JAK3 –/– mice were considerably reduced in size (between 0.5% and 10% of the normal organ) although the cell composition was essentially normal. Interestingly, a failure of T cells to respond to a proliferative signal (phorbol 12-myristate 13-acetate [PMA] plus ionomycin), and a profound diminution of T cells to produce IL-2 in response to anti-CD3 antibodies plus anti-CD28 antibodies,[466] demonstrated an important role for JAK3 in T cell biology.

The similarity of the B cell phenotype of JAK3 –/– mice to mice in which the γ_c has been deleted[458] and those in which the IL-7 receptor α-chain has been disrupted[468,469] confirm the notion that these three genes are crucial components of the same pathway functioning in lymphoid development. Taken together, these studies have cast important light upon the role of JAK3 in lymphoid development, and although none of the components of the phenotype were surprising, they nonetheless confirm the expected role of JAK3.

CLOSING REMARKS

An impressive amount of information has been gathered on piecing together the mechanics of the JAK-STAT pathway since the discovery of the participant proteins. Already the bare bones of this pathway are documented, and more recently some nuances and intricacies are beginning to emerge. Naturally, with our current level of knowledge, surprising new wrinkles are discovered more frequently than entirely expected results. Witness, for example, the STAT1 knock-out mice,[475] which, despite lacking a molecule which is demonstrably recruited by more cytokines than any other, is nonetheless, indispensable only for interferon expression, the pathway in which it was discovered! There remain a number of pieces of unfinished business in this signal transduction pathway, the resolution of which is bound to lead to new and interesting concepts in signal transduction in general and the JAK-STAT pathway in particular.

For example, what is the role of the STAT transcription factor complex in the activation of gene expression? Since the activation of the STATs appears to be transient (at least as measured by tyrosine phosphorylation and EMSA analysis), why is it that activation of the genes regulated by these complexes remains high well after the dissipation of the activated STAT complex? Indeed, what is it that switches the STAT off? Is there a nuclear phosphotyrosine phosphatase which extinguishes the signal? Are there other components of STAT complexes which are yet to be discovered? The demonstration that p48 is a key component of ISGF3, and the more recent data showing that JUN interacts with a variant of STAT3, seem to suggest that heteromeric complexes are the norm rather than the exception in this system.

In chapter 7, we presented the notion that the assembly of a STAT complex is entirely dependent upon the tyrosine phosphorylation of the Y701 sites at the C-terminal tail of the STAT. Indeed, there is strong evidence that this is the key event in the formation of a STAT containing transcription complex. Recent data has hinted that this is not the whole story. During a search for evidence for the association of members of the STAT family with Hsp90 (an attempt to solve another of the mysteries of the JAK-STAT pathway; how do the activated STAT complexes move into the nucleus?), it was discovered that under mild extraction conditions STAT1/STAT3 dimers can be found which completely lack any phosphotyrosine.[470] Naturally this is a worrying result to those who favor the phosphotyrosine/SH2 nexus as the most important level of control of all signal transduction processes! On balance, it does not seem improbable that the interaction of the two STATs in a heterodimer relies upon protein surfaces other than those offered by the SH2/phosphotyrosine interaction, and it may be that these preformed complexes are a low level manifestation of the availability of these surfaces to interact with each other, even in the absence of a cytokine receptor mediated recruitment signal. Alternatively, the preformed complexes may represent the true form of latent cytoplasmic STAT, and the cytokine recruitment of STATs to the receptor may produce a mere tweaking of the structure (via the SH2/phosphotyrosine interaction) into an "active" conformation.

More challenges to the neat model of the JAK-STAT pathway are being faced on the JAK front. The IFNγ pathway was a staple for the somatic cell genetic screen described in chapter 6. Here, the absolute dependence of the pathway upon JAK1, JAK2 and STAT1 was shown with considerable elegance. Recent work has suggested that we still have a lot to learn about the roles of each of these components in this pathway.[471] Using the same IFNα/β and IFNγ pathway defective mutant cell line as in earlier studies, (U4A[300]), it was shown that kinase defective mutants of JAK1 were able to restore the induction of IFNγ-inducible genes. Moreover, enzymatically active JAK1 was not required for the phosphorylation/activation of JAK2, or the activation of STAT1 (albeit at lower levels than in the presence of fully functional JAK1 protein). The IFNα/β and IL-6 pathways were unable to be complemented by this defective JAK1 mutant, implying a different role for this protein in these different signal transduction pathways. Interestingly, while gene induction appeared to be restored by a kinase defective JAK1, the ability to mount an effective anti-viral response by these cells was not restored, suggesting an additional (and largely unanticipated) layer of complexity. Taken together, these data suggest that JAK1 may play several roles in the activation of a given signal transduction pathway. Firstly, the PTK activity of JAK1 appears to be a necessary feature of IFNα/β and IL-6 signaling.[472] Secondly, it may be that JAK1 is required for the generation of a stable receptor complex. Thirdly, JAK1 may be part of a docking site for a STAT. This possibility has been canvassed in other work on the IFNα/β system.[473] Here, the demonstration of an association of JAK1 and Tyk2 with the SH2 domains of STAT1 and STAT2, hints that there may be a role for each of these molecules in the marshaling of the STATs into activated transcription complexes. Thus, the role of JAK1 in offering a PTK domain to the IFNα/β receptor complex may be replaced by the JAK2 PTK domain in JAK1 kinase-defective transfected cells. While its other role as a docking site for STATs may not require its PTK activity, this is perhaps not something that JAK2 can replace in the pathway. On balance it seems likely that the dual role of JAK1 in IFNα/β signaling is telling us something important, although what this is remains enigmatic.

As discussed above, the integration of the JAK-STAT pathway into the overall fabric of signal transduction is beginning to emerge as a key theme for future research in this area. Are there JAK mediated phenomena which do not involve STATs? Similarly, are there inputs into the STATs which are not mediated by the JAK family of PTKs? Finally, how are the multiple steady-state interactions of the wide range of different cytokine molecules which bombard each and every cell in the body, integrated by the intracellular circuitry? How, for instance would a T cell integrate the simultaneous triggering of the IL-2 receptor and the IFNγ receptor? The answer to these questions will lead, ultimately, to the integration of all of the various *pathways* into a *network*. These are exciting times.

REFERENCES

1. Davis S, Aldrich TH, Valenzuela DM et al. The receptor for ciliary neurotrophic factor. Science 1991; 253:59-63.
2. Hunter T. A thousand and one protein kinases. Cell 1987; 50:823-9.
3. Hunter T. Protein kinase classification. Methods Enzymol 1991; 200:3-37.
4. Hanks SK, Quinn AM, Hunter T. The protein kinase family: conserved features and deduced phylogeny of the catalytic domains. Science 1988; 241:42-52.
5. Hanks SK, Quinn AM. Protein kinase catalytic domain sequence data-base: identification of conserved features of primary structure and classification of family members. Methods Enzymol 1991; 200:38-62.
6. Hanks SK, Quinn AM. Protein kinase catalytic domain sequence database: identification of conserved features of primary structure and classification of family members. Methods Enzymol 1991; 200:38-62.
7. Hardie G, Hanks S. The Protein Kinase Facts Book. 1996.
8. Songyang Z, Blechner S, Hoagland N et al. Use of an oriented peptide library to determine the optimal substrates of protein kinases. Curr Biol 1994; 4:973-82.
9. Ullrich A, Schlessinger J. Signal transduction by receptors with tyrosine kinase activity. Cell 1990; 61:203-12.
10. Bauskin AR, Alkalay I, Ben Neriah Y. Redox regulation of a protein tyrosine kinase in the endoplasmic reticulum. Cell 1991; 66:685-96.
11. Bishop JM. Molecular themes in oncogenesis. Cell 1991; 64:235-48.
12. Hunter T. Cooperation between oncogenes. Cell 1991; 64:249-70.
13. Songyang Z, Shoelson SE, Chaudhuri M et al. SH2 domains recognize specific phosphopeptide sequences. Cell 1993; 72:762-78.
14. Songyang Z, Shoelson SE, McGlade J et al. Specific motifs recognized by the SH2 domains of Csk, 3BP2, fps/fes, GRB-2, HCP, SHC, Syk, and Vav. Mol Cell Biol 1994; 14:2777-85.
15. Koch CA, Anderson D, Moran MF et al. SH2 and SH3 domains: elements that control interactions of cytoplasmic signaling proteins. Science 1991; 252:668-74.

16. Anderson D, Koch CA, Grey L et al. Binding of SH2 domains of phospholipase C gamma 1, GAP, and Src to activated growth factor receptors. Science 1990; 250:979-82.
17. McGlade CJ, Ellis C, Reedijk M et al. SH2 domains of the p85 alpha subunit of phosphatidylinositol 3-kinase regulate binding to growth factor receptors. Mol Cell Biol 1992; 12:991-7.
18. Sadowski I, Stone JC, Pawson T. A noncatalytic domain conserved among cytoplasmic protein-tyrosine kinases modifies the kinase function and transforming activity of Fujinami Sarcoma Virus $P130^{gag\text{-}fps}$. Mol Cell Biol 1986; 6:4396-408.
19. Kitamura N, Yoshida M. Small deletions in *src* of Rous sarcoma virus modifying transformation phenotypes: identification of 207-nucleotide deletion and its smaller product with protein kinase activity. J Virol 1983; 46:985-92.
20. Stone JC, Atkinson T, Smith M et al. Identification of functional regions in the transforming proteins of Fujinami sarcoma virus by in-phase insertion mutagenesis. Cell 1984; 37:549-58.
21. Katzav S, Martin Zanca D, Barbacid M. *vav*, a novel human oncogene derived from a locus ubiquitously expressed in hematopoietic cells. EMBO J 1989; 8:2283-90.
22. Margolis B, Hu P, Katzav S et al. Tyrosine phosphorylation of vav proto-oncogene product containing SH2 domain and transcription factor motifs [see comments]. Nature 1992; 356:71-4.
23. Lehmann JM, Riethmuller G, Johnson JP. Nck, a melanoma cDNA encoding a cytoplasmic protein consisting of the src homology units SH2 and SH3. Nucleic Acids Res 1990; 18:1048.
24. Lowenstein EJ, Daly RJ, Batzer AG et al. The SH2 and SH3 domain-containing protein GRB2 links receptor tyrosine kinases to ras signaling. Cell 1992; 70:431-42.
25. Clark SG, Stern MJ, Horvitz HR. C. elegans cell-signaling gene sem-5 encodes a protein with SH2 and SH3 domains [see comments]. Nature 1992; 356:340-4.
26. Skolnik EY, Margolis B, Mohammadi M et al. Cloning of PI3 kinase-associated p85 utilizing a novel method for expression/cloning of target proteins for receptor tyrosine kinases. Cell 1991; 65:83-90.
27. Songyang Z, Shoelson SE, McGlade J et al. Specific motifs recognized by the SH2 domains of Csk, 3BP2, fps/fes, GRB-2, HCP, SHC, Syk, and Vav. Mol Cell Biol 1994; 14:2777-85.
28. Cantley LC, Auger KR, Carpenter C et al. Oncogenes and signal transduction [published erratum appears in Cell 1991 May 31;65(5):following 914]. Cell 1991; 64:281-302.
29. Cantley LC, Songyang Z. Specificity in recognition of phosphopeptides by src-homology 2 domains. J Cell Sci Suppl 1994; 18:121-6.

30. Overduin M, Rios CB, Mayer BJ et al. Three-dimensional solution structure of the src homology domain of c-abl. Cell 1992; 70:697-704.
31. Overduin M, Mayer B, Rios CB et al. Secondary structure of Src homology 2 domain of c-Abl by heteronuclear NMR spectroscopy in solution. Proc Natl Acad Sci USA 1992; 89:11673-7.
32. Eck MJ, Shoelson SE, Harrison SC. Recognition of a high-affinity phosphotyrosyl peptide by the Src homology-2 domain of p56lck. Nature 1993; 362:87-91.
33. Eck MJ, Atwell SK, Shoelson SE et al. Structure of the regulatory domains of the Src-family tyrosine kinase Lck. Nature 1994; 368:764-9.
34. Waksman G, Kominos D, Robertson SC et al. Crystal structure of the phosphotyrosine recognition domain SH2 of v-src complexed with tyrosine-phosphorylated peptides [see comments]. Nature 1992; 358:646-53.
35. Waksman G, Shoelson SE, Pant N et al. Binding of a high affinity phosphotyrosyl peptide to the Src SH2 domain:crystal structures of the complexed and peptide-free forms. Cell 1993; 72:779-90.
36. Superti Furga G, Fumagalli S, Koegl M et al. Csk inhibition of c-Src activity requires both the SH2 and SH3 domains of Src. EMBO J 1993; 12:2625-34.
37. Ren R, Mayer BJ, Cicchetti P et al. Identification of a ten-amino acid proline-rich SH3 binding site. Science 1993; 259:1157-61.
38. Panayotou G, Waterfield MD. The assembly of signaling complexes by receptor tyrosine kinases. Bioessays 1993; 15:171-7.
39. Rozengurt E. Early signals in the mitogenic response. Science 1986; 234:161-6.
40. Heldin CH, Backstrom G, Ostman A et al. Binding of different dimeric forms of PDGF to human fibroblasts: evidence for two separate receptor types. EMBO J 1988; 7:1387-93.
41. Cunningham BC, Ultsch M, De Vos AM et al. Dimerization of the extracellular domain of the human growth hormone receptor by a single hormone molecule. Science 1991; 254:821-5.
42. Schlessinger J, Ullrich A. Growth factor signaling by receptor tyrosine kinases. Neuron 1992; 9:383-91.
43. Rozakis Adcock M, Fernley R, Wade J et al. The SH2 and SH3 domains of mammalian Grb2 couple the EGF receptor to the Ras activator mSos1 [see comments]. Nature 1993; 363:83-5.
44. Buday L, Downward J. Epidermal growth factor regulates p21ras through the formation of a complex of receptor, Grb2 adaptor protein and Sos nucleotide exchange factor. Cell 1993; 73:611-20.
45. Skolnik EY, Batzer A, Li N et al. The function of GRB2 in linking the insulin receptor to Ras signaling pathways. Science 1993; 260:1953-5.

46. Baltensperger K, Kozma LM, Cherniack AD et al. Binding of the Ras activator son of sevenless to insulin receptor substrate-1 signaling complexes. Science 1993; 260:1950-2.
47. Skolnik EY, Lee CH, Batzer A et al. The SH2/SH3 domain-containing protein GRB2 interacts with tyrosine-phosphorylated IRS1 and Shc: implications for insulin control of ras signaling. EMBO J 1993; 12:1929-36.
48. Egan SE, Giddings BW, Brooks MW et al. Association of Sos Ras exchange protein with Grb2 is implicated in tyrosine kinase signal transduction and transformation [see comments]. Nature 1993; 363:45-51.
49. Kazlauskas A. Receptor tyrosine kinases and their targets. Curr Opin Genet Dev 1994; 4:5-14.
50. Barbacid M. *ras* genes. Ann Rev Biochem 1987; 56:779-827.
51. Dickson B, Hafen E. Genetics of signal transduction in invertebrates. Curr Opin Genet Dev 1994; 4:64-70.
52. Smith MR, DeGudicibus SJ, Stacey DW. Requirement for c-*ras* proteins during viral oncogene transformation. Nature 1986; 320:540-3.
53. McCormick F. Activators and effectors of *ras* p21 proteins. Curr Opin Genet Dev 1994; 4:71-6.
54. Marshall CJ. MAP kinase kinase kinase, MAP kinase kinase and MAP kinase. Curr Opin Genet Dev 1994; 4:82-9.
55. Davis RJ. The mitogen-activated protein kinase signal transduction pathway. J Biol Chem 1993; 268:14553-6.
56. Moodie SA, Willumsen BM, Weber MJ et al. Complexes of Ras.GTP with Raf-1 and mitogen-activated protein kinase kinase [see comments]. Science 1993; 260:1658-61.
57. van Aelst L, Barr M, Marcus S et al. Complex formation between RAS and RAF and other protein kinases. Proc Natl Acad Sci USA 1993; 90:6213-7.
58. Vojtek AB, Hollenberg SM, Cooper JA. Mammalian Ras interacts directly with the serine/threonine kinase Raf. Cell 1993; 74:205-14.
59. Zhang XF, Settleman J, Kyriakis JM et al. Normal and oncogenic p21ras proteins bind to the amino-terminal regulatory domain of c-Raf-1. Nature 1993; 364:308-13.
60. Warne PH, Viciana PR, Downward J. Direct interaction of Ras and the amino-terminal region of Raf-1 in vitro. Nature 1993; 364:352-5.
61. Williams NG, Roberts TM, Li P. Both p21ras and pp60v-src are required, but neither alone is sufficient, to activate the Raf-1 kinase. Proc Natl Acad Sci USA 1992; 89:2922-6.
62. Turner BC, Tonks NK, Rapp UR et al. Interleukin 2 regulates Raf-1 kinase activity through a tyrosine phosphorylation-dependent

mechanism in a T-cell line. Proc Natl Acad Sci USA 1993; 90:5544-8.

63. Fantl WJ, Escobedo JA, Martin GA et al. Distinct phosphotyrosines on a growth factor receptor bind to specific molecules that mediate different signaling pathways. Cell 1992; 69:413-23.

64. Backer JM, Myers MG Jr, Shoelson SE et al. Phosphatidylinositol 3'-kinase is activated by association with IRS-1 during insulin stimulation. EMBO J 1992; 11:3469-79.

65. Prasad KV, Cai YC, Raab M et al. T-cell antigen CD28 interacts with the lipid kinase phosphatidylinositol 3-kinase by a cytoplasmic Tyr(P)-Met-Xaa-Met motif. Proc Natl Acad Sci USA 1994; 91:2834-8.

66. Reedijk M, Liu X, van der Geer P et al. Tyr721 regulates specific binding of the CSF-1 receptor kinase insert to PI 3'-kinase SH2 domains: a model for SH2-mediated receptor-target interactions. EMBO J 1992; 11:1365-72.

67. Lane HA, Fernandez A, Lamb NJ et al. p70s6k function is essential for G1 progression. Nature 1993; 363:170-2.

68. Berridge MJ. Inositol trisphosphate and calcium signaling. Nature 1993; 361:315-25.

69. Nishizuka Y. Intracellular signaling by hydrolysis of phospholipids and activation of Protein Kinase C. Science 1992; 258:607-14.

70. Rotin D, Margolis B, Mohammadi M et al. SH2 domains prevent tyrosine dephosphorylation of the EGF receptor: identification of Tyr992 as the high-affinity binding site for SH2 domains of phospholipase C gamma. EMBO J 1992; 11:559-67.

71. Mohammadi M, Honegger AM, Rotin D et al. A tyrosine-phosphorylated carboxy-terminal peptide of the fibroblast growth factor receptor (Flg) is a binding site for the SH2 domain of phospholipase C-gamma 1. Mol Cell Biol 1991; 11:5068-78.

72. Pascal SM, Singer AU, Gish G et al. Nuclear magnetic resonance structure of an SH2 domain of phospholipase C-gamma 1 complexed with a high affinity binding peptide. Cell 1994; 77:461-72.

73. Farrow NA, Muhandiram R, Singer AU et al. Backbone dynamics of a free and phosphopeptide-complexed Src homology 2 domain studied by 15N NMR relaxation. Biochemistry 1994; 33:5984-6003.

74. Myers MG, Jr., Sun XJ, White MF. The IRS-1 signaling system. TIBS 1994; 19:289-93.

75. White MF, Maron R, Kahn CR. Insulin rapidly stimulates tyrosine phosphorylation of a Mr-185,000 protein in intact cells. Nature 1985; 318:183-6.

76. Sun XJ, Rothenberg P, Kahn CR et al. Structure of the insulin receptor substrate IRS-1 defines a unique signal transduction protein. Nature 1991; 352:73-7.

77. Sun XJ, Crimmins DL, Myers MG Jr et al. Pleiotropic insulin signals are engaged by multisite phosphorylation of IRS-1. Mol Cell Biol 1993; 13:7418-28.
78. Wang LM, Keegan AD, Li W et al. Common elements in interleukin 4 and insulin signaling pathways in factor-dependent hematopoietic cells. Proc Natl Acad Sci USA 1993; 90:4032-6.
79. Bazan JF. Structural design and molecular evolution of a cytokine receptor superfamily. Proc Natl Acad Sci USA 1990; 87:6934-8.
80. Bazan JF. Shared architecture of hormone binding domains in type I and II interferon receptors [letter]. Cell 1990; 61:753-4.
81. Bazan JF. Haemopoietic receptors and helical cytokines. Immunol Today 1990; 11:350-4.
82. Ihle JN, Keller J, Oroszlan S et al. Biologic properties of homogeneous interleukin 3. I. Demonstration of WEHI-3 growth factor activity, mast cell growth factor activity, p cell-stimulating factor activity, colony-stimulating factor activity, and histamine-producing cell-stimulating factor activity. J Immunol 1983; 131:282-7.
83. Gough NM, Gough J, Metcalf D et al. Molecular cloning of cDNA encoding a murine haematopoietic growth regulator, granulocyte-macrophage colony stimulating factor. Nature 1984; 309:763-7.
84. Gough NM, Metcalf D, Gough J et al. Structure and expression of the mRNA for murine granulocyte-macrophage colony stimulating factor. EMBO J 1985; 4:645-53.
85. Jacobs K, Shoemaker C, Rudersdorf R et al. Isolation and characterization of genomic and cDNA clones of human erythropoietin. Nature 1985; 313:806-10.
86. Souza LM, Boone TC, Gabrilove J et al. Recombinant human granulocyte colony-stimulating factor: effects on normal and leukemic myeloid cells. Science 1986; 232:61-5.
87. Lin FK, Suggs S, Lin CH et al. Cloning and expression of the human erythropoietin gene. Proc Natl Acad Sci USA 1985; 82:7580-4.
88. Kaushansky K, Lok S, Holly RD et al. Promotion of megakaryocyte progenitor expansion and differentiation by the c-Mpl ligand thrombopoietin [see comments]. Nature 1994; 369:568-71.
89. Lok S, Kaushansky K, Holly RD et al. Cloning and expression of murine thrombopoietin cDNA and stimulation of platelet production in vivo [see comments]. Nature 1994; 369:565-8.
90. Robb RJ, Kutny RM, Chowdhry V. Purification and partial sequence analysis of human T-cell growth factor. Proc Natl Acad Sci USA 1983; 80:5990-4.
91. Noma Y, Sideras P, Naito T et al. Cloning of cDNA encoding the murine IgG1 induction factor by a novel strategy using SP6 promoter. Nature 1986; 319:640-6.

92. Yokota T, Otsuka T, Mosmann T et al. Isolation and characterization of a human interleukin cDNA clone, homologous to mouse B-cell stimulatory factor 1, that expresses B-cell- and T-cell-stimulating activities. Proc Natl Acad Sci USA 1986; 83:5894-8.
93. Namen AE, Lupton S, Hjerrild K et al. Stimulation of B-cell progenitors by cloned murine interleukin-7. Nature 1988; 333:571-3.
94. Van Snick J, Goethals A, Renauld JC et al. Cloning and characterization of a cDNA for a new mouse T cell growth factor (P40). J Exp Med 1989; 169:363-8.
95. Hirano T, Yasukawa K, Harada H et al. Complementary DNA for a novel human interleukin (BSF-2) that induces B lymphocytes to produce immunoglobulin. Nature 1986; 324:73-6.
96. Gough NM, Williams RL. The pleiotropic actions of Leukemia Inhibitory Factor. Cancer Cells 1989; 1:77-80.
97. Stockli KA, Lottspeich F, Sendtner M et al. Molecular cloning, expression and regional distribution of rat ciliary neurotrophic factor. Nature 1989; 342:920-3.
98. Gray PW, Leung DW, Pennica D et al. Expression of human immune interferon cDNA in E. coli and monkey cells. Nature 1982; 295:503-8.
99. Pestka S. The purification and manufacture of human interferons. Their promise is still not fulfilled, but now their genes have been isolated and cloned in E. coli. The bacteria are making interferon in quantity and the bacterial product is undergoing clinical trials. Sci Am 1983; 249:37-43.
100. Pestka S. The human interferons—from protein purification and sequence to cloning and expression in bacteria: before, between, and beyond. Arch Biochem Biophys 1983; 221:1-37.
101. Moore KW, Vieira P, Fiorentino DF et al. Homology of cytokine synthesis inhibitory factor (IL-10) to the Epstein-Barr virus gene BCRFI [published erratum appears in Science 1990 Oct 26; 250(4980):494]. Science 1990; 248:1230-4.
102. De Vos AM, Ultsch M, Kossiakoff AA. Human growth hormone and extracellular domain of its receptor: crystal structure of the complex. Science 1992; 255:306-12.
103. Fukunaga R, Ishizaka Ikeda E, Pan C-X et al. Functional domains of the granulocyte colony-stimulating factor receptor. EMBO 1991; 10:2855-65.
104. VanderKuur JA, Wang X, Zhang L et al. Domains of the growth hormone receptor required for association and activation of JAK2 tyrosine kinase. J Biol Chem 1994; 269:21709-17.
105. Frank SJ, Gilliland G, Kraft AS et al. Interaction of the growth hormone receptor cytoplasmic domain with the JAK2 tyrosine kinase. Endocrinology 1994; 135:2228-39.

106. Fukunaga R, Ishizaka-Ikeda E, Nagata S. Growth and differentiation signals mediated by different regions in the cytoplasmic domain of granulocyte colony-stimulating factor receptor. Cell 1993; 74:1079-87.
107. Miyazaki T, Kawahara A, Fujii H et al. Functional activation of Jakl and Jak3 by selective association with IL-2 receptor subunits. Science 1994; 266:1045-7.
108. Damen JE, Liu L, Cutler RL et al. Erythropoietin stimulates the tyrosine phosphorylation of Shc and its association with Grb2 and a 145-Kd tyrosine phosphorylated protein. Blood 1993; 82: 2296-303.
109. Isfort RJ, Ihle JN. Multiple hematopoietic growth factors signal through tyrosine phosphorylation. Growth Factors 1990; 2:213-20.
110. Murakami M, Narazaki M, Hibi M et al. Critical cytoplasmic region of the interleukin 6 signal transducer gp130 is conserved in the cytokine receptor family. Proc Natl Acad Sci USA 1991; 88:11349-53.
111. Songyang Z, Shoelson SE, McGlade J et al. Specific motifs recognized by the SH2 domains of Csk, 3BP2, fps/fes, GRB-2, HCP, SHC, Syk, and Vav. Mol Cell Biol 1994; 14:2777-85.
112. Stahl N, Farruggella TJ, Boulton TG et al. Choice of STATs and other substrates specified by modular tyrosine-based motifs in cytokine receptors. Science 1995; 267:1349-53.
113. Lebrun JJ, Ali S, Goffin V et al. A single phosphotyrosine residue of the prolactin receptor is responsible for activation of gene transcription. Proc Natl Acad Sci USA 1995; 92:4031-5.
114. Damen JE, Cutler RL, Jiao H et al. Phosphorylation of tyrosine 503 in the erythropoietin receptor (EpR) is essential for binding the P85 subunit of phosphatidylinositol (PI) 3-kinase and for EpR-associated PI 3-kinase activity. J Biol Chem 1995; 270:23402-8.
115. Yan H, Krishnan K, Greenlund AC et al. Phosphorylated interferon-α receptor 1 subunit (IFNaR1) acts as a docking site for the latent form of the 113 kDa STAT2 protein. EMBO Journal 1996; 15:1064-74.
116. Ihle JN. Cytokine receptor signaling. Nature 1995; 377:591-4.
117. Fukunaka R, Ishizaka-Ikeda E, Seto Y, and Nagata S. Expression cloning of a receptor for murine granulocyte colony stimulating factor. Cell 1990; 61341-50.
118. Yamasaki K, Taga T, Hirata Y et al. Cloning and expression of the human interleukin-6 (BSF-2/IFN beta 2) receptor. Science 1988; 241:825-8.
119. Taga T, Hibi M, Hirata Y et al. Interleukin-6 triggers the association of its receptor with a possible signal transducer, gp130. Cell 1989; 58:573-81.

120. Gearing DP, Thut CJ, VandenBos T et al. Leukemia inhibitory factor receptor is structurally related to the IL-6 signal transducer, gp130. EMBO 1991; 10:2839-48.
121. Hibi M, Murakami M, Saito M et al. Molecular cloning and expression of an IL-6 signal transducer, gp130. Cell 1990; 63:1149-57.
122. Cook JR, Jung V, Schwartz B et al. Structural analysis of the human interferon γ receptor: A small segment of the intracellular domain is specifically required for class I major histocompatibility complex antigen induction and antiviral activity. Proc Natl Acad Sci USA 1992; 89:11317-21.
123. Stahl N, Yancopoulos GD. The tripartite CNTF receptor complex: activation and signaling involves components shared with other cytokines. J Neurobiol 1994; 25:1454-66.
124. Hilton DJ, Hilton AA, Raicevic A et al. Cloning of a murine IL-11 receptor alpha-chain; requirement for gp130 for high affinity binding and signal transduction. EMBO J 1994; 13:4765-75.
125. Stahl N, Yancopoulos GD. The alphas, betas, and kinases of cytokine receptor complexes. Cell 1993; 74:587-90.
126. Gearing DP, Comeau MR, Friend DJ et al. The IL-6 signal transducer, gp130: an oncostatin M receptor and affinity converter for the LIF receptor. Science 1992; 255:1434-7.
127. Gearing DP, King JA, Gough NM et al. Expression cloning of a receptor for human granulocyte-macrophage colony-stimulating factor. EMBO 1993; 8:3667-76.
128. Neuhaus H, Bettenhausen B, Bilinski P et al. Etl2, a novel putative type-I cytokine receptor expressed during mouse embryogenesis at high levels in skin and cells with skeletogenic potential. Dev Biol 1994; 166:531-42.
129. Murata Y, Takaki S, Migita M et al. Molecular cloning and expression of the human interleukin 5 receptor. J Exp Med 1992; 175:341-51.
130. Gorman DM, Itoh N, Kitamura T et al. Cloning and expression of a gene encoding an interleukin 3 receptor-like protein: identification of another member of the cytokine receptor gene family. Proc Natl Acad Sci USA 1990; 87:5459-63.
131. Itoh N, Yonehara S, Schreurs J et al. Cloning of an interleukin-3 receptor gene: a member of a distinct receptor gene family. Science 1990; 247:324-7.
132. Leonard WJ, Depper JM, Crabtree GR et al. Molecular cloning and expression of cDNAs for the human interleukin-2 receptor. Nature 1984; 311:626-31.
133. Hatakeyama M, Tsudo M, Minamoto S et al. Interleukin-2 receptor beta chain gene: generation of three receptor forms by cloned human alpha and beta chain cDNA's. Science 1989; 244:551-6.

134. Takeshita T, Asao H, Ohtani K et al. Cloning of the gamma chain of the human IL-2 receptor. Science 1992; 257:379-82.
135. Mosley B, Beckmann MP, March CJ et al. The murine interleukin-4 receptor: molecular cloning and characterization of secreted and membrane bound forms. Cell 1989; 59:335-48.
136. Goodwin RG, Friend D, Ziegler SF et al. Cloning of the human and murine interleukin-7 receptors: demonstration of a soluble form and homology to a new receptor superfamily. Cell 1990; 60:941-51.
137. Renauld JC, Druez C, Kermouni A et al. Expression cloning of the murine and human interleukin 9 receptor cDNAs. Proc Natl Acad Sci USA 1992; 89:5690-4.
138. Vita N, Lefort S, Laurent P et al. Characterization and comparison of the interleukin 13 receptor with the interleukin 4 receptor on several cell types. J Biol Chem 1995; 270:3512-7.
139. Giri JG, Kumaki S, Ahdieh M et al. Identification and cloning of a novel IL-15 binding protein that is structurally related to the alpha chain of the IL-2 receptor. EMBO J 1995; 14:3654-63.
140. Kondo M, Takeshita T, Ishii N et al. Sharing of the interleukin-2 (IL-2) receptor gamma chain between receptors for IL-2 and IL-4. Science 1993; 262:1874-7.
141. Noguchi M, Nakamura Y, Russell SM et al. Interleukin-2 receptor gamma chain: a functional component of the interleukin-7 receptor. Science 1994; 262:1877-83.
142. Kondo M, Takeshita T, Higuchi M et al. Functional participation of the IL-2 receptor gamma chain in IL-7 receptor complexes. Science 1994; 263:1453-4.
143. Giri JG, Ahdieh M, Eisenman J et al. Utilization of the beta and gamma chains of the IL-2 receptor by the novel cytokine IL-15. EMBO J 1994; 13:2822-30.
144. Obiri NI, Debinski W, Leonard WJ et al. Receptor for interleukin 13. Interaction with interleukin 4 by a mechanism that does not involve the common gamma chain shared by receptors for interleukins 2, 4, 7, 9, and 15. J Biol Chem 1995; 270:8797-804.
145. Kawahara A, Minami Y, Taniguchi T. Evidence for a critical role for the cytoplasmic region of the interleukin 2 (IL-2) receptor gamma chain in IL-2, IL-4, and IL-7 signaling. Mol Cell Biol 1994; 14:5433-40.
146. Lin JX, Migone TS, Tsang M et al. The role of shared receptor motifs and common Stat proteins in the generation of cytokine pleiotropy and redundancy by IL-2, IL-4, IL-7, IL-13, and IL-15. Immunity 1995; 2:331-9.
147. Morella KK, Lai CF, Kumaki S et al. The action of interleukin-2 receptor subunits defines a new type of signaling mechanism for hematopoietin receptors in hepatic cells and fibroblasts. J Biol Chem 1995; 270:8298-310.

148. Leonard WJ, Noguchi M, Russell SM et al. The molecular basis of X-linked severe combined immunodeficiency: the role of the interleukin-2 receptor gamma chain as a common gamma chain, gamma c. Immunol Rev 1994; 138:61-86.
149. Russell SM, Johnston JA, Noguchi M et al. Interaction of IL-2Rβ and gamma chains with Jak1 and Jak3: implications for XSCID and XCID. Science 1994; 266:1042-5.
150. Uze G, Lutfalla G, Gresser I. Genetic transfer of a functional human interferon α receptor into mouse cells: cloning and expression of its cDNA. Cell 1990; 60:225-34.
151. Novick D, Cohen B, Rubinstein M. The human interferon α/β receptor:characterization and molecular cloning. Cell 1994; 77:391-400.
152. Aguet M, Dembic Z, Merlin G. Molecular cloning and expression of the human interferon-gamma receptor. Cell 1988; 55:273-80.
153. Mahoney SE, Duvic M, Nickoloff BJ et al. Human immunodeficiency virus (HIV) transcripts identified in HIV-related psoriasis and Kaposi's sarcoma lesions. J Clin Invest 1991; 88:174-85.
154. Soh J, Donnelly RJ, Kotenko S et al. Identification and sequence of an accessory factor required for activation of the human interferon gamma receptor. Cell 1994; 76:793-802.
155. Ho AS, Liu Y, Khan TA et al. A receptor for interleukin 10 is related to interferon receptors. Proc Natl Acad Sci USA 1993; 90:11267-71.
156. Lutfalla G, Holland SJ, Cinato E et al. Mutant U5A cells are complemented by an interferon-alpha beta receptor subunit generated by alternative processing of a new member of a cytokine receptor gene cluster. EMBO J 1995; 14:5100-8.
157. Liu Y, Wei SH, Ho AS et al. Expression cloning and characterization of a human IL-10 receptor. J Immunol 1994; 152:1821-9.
158. Wang LM, Keegan AD, Paul WE et al. IL-4 activates a distinct signal transduction cascade from IL-3 in factor-dependent myeloid cells. EMBO J 1992; 11:4899-908.
159. Satoh T, Nakafuku M, Miyajima A et al. Involvement of ras p21 protein in signal-transduction pathways from interleukin 2, interleukin 3, and granulocyte/macrophage colony-stimulating factor, but not from interleukin 4. Proc Natl Acad Sci USA 1991; 88:3314-8.
160. Satoh T, Uehara Y, Kaziro Y. Inhibition of interleukin 3 and granulocyte-macrophage colony-stimulating factor stimulated increase of active ras.GTP by herbimycin A, a specific inhibitor of tyrosine kinases. J Biol Chem 1992; 267:2537-41.
161. Duronio V, Welham MJ, Abraham S et al. p21ras activation via hemopoietin receptors and c-kit requires tyrosine kinase activity but

not tyrosine phosphorylation of p21ras GTPase-activating protein. Proc Natl Acad Sci USA 1992; 89:1587-91.
162. Torti M, Bencke Marti K, Altschuler D et al. Erythropoietin induces p21^{ras} activation and p120GAP tyrosine phosphorylation in human erythroleukemia cells. J Biol Chem 1992; 267:8293-8.
163. Welham MJ, Duronio V, Sanghera JS et al. Multiple hemopoietic growth factors stimulate activation of mitogen-activated protein kinase family members. J Immunol 1992; 149:1683-93.
164. Wilks AF, Kurban RR. Isolation and structural analysis of murine c-fes cDNA clones. Oncogene 1988; 3:289-94.
165. Wilks AF. Two putative protein-tyrosine kinases identified by application of the polymerase chain reaction. Proc Natl Acad Sci USA 1989; 86:1603-7.
166. Harpur AG, Andres AC, Ziemiecki A et al. JAK2, a third member of the JAK family of protein tyrosine kinases. Oncogene 1992; 7:1347-53.
167. Wilks AF, Harpur AG, Kurban RR et al. Two novel protein-tyrosine kinases, each with a second phosphotransferase-related catalytic domain, define a new class of protein kinase. Mol Cell Biol 1991; 11:2057-65.
168. Krolewski JJ, Lee R, Eddy R et al. Identification and chromosomal mapping of new human tyrosine kinase genes. Oncogene 1990; 5:277-82.
169. Firmbach-Kraft I, Byers M, Shows T et al. Tyk2, prototype of a novel class of non-receptor tyrosine kinase genes. Oncogene 1990; 5:1329-36.
170. Wilks AF, Kurban RR, Hovens CM et al. The application of the polymerase chain reaction to cloning members of the protein tyrosine kinase family. Gene 1989; 85:67-74.
171. Wilks AF. Cloning members of protein-tyrosine kinase family using polymerase chain reaction. Methods Enzymol 1991; 200:533-46.
172. Anonymous. Encyclopedia Britannica. 11th ed. Janus. 1989:155-6.
173. Kawamura M, McVicar DW, Johnston JA et al. Molecular cloning of L-JAK, a Janus family protein-tyrosine kinase expressed in natural killer cells and activated leukocytes. Proc Natl Acad Sci USA 1994; 91:6374-8.
174. Rane SG, Reddy EP. JAK3: a novel JAK kinase associated with terminal differentiation of hematopoietic cells. Oncogene 1994; 9:2415-23.
175. Witthuhn BA, Silvennoinen O, Miura O et al. Involvement of the Jak-3 Janus kinase in signaling by interleukins 2 and 4 in lymphoid and myeloid cells. Nature 1994; 370:153-7.
176. Takahashi T, Shirasawa T. Molecular cloning of rat JAK3, a novel member of the JAK family of protein tyrosine kinases. FEBS Lett 1994; 342:124-8.

177. Binari R, Perrimon N. Stripe specific regulation of pair-rule genes by hopscotch, a putative JAK family tyrosine kinase. Genes Dev 1994; 8:300-12.
178. Kozak M. Compilation and analysis of sequences upstream from the translational start site in eukaryotic mRNAs. Nucl Acid Res 1984; 12:857-72.
179. Yang X, Chung D, Cepko CL. Molecular cloning of the murine JAK1 protein tyrosine kinase and its expression in the mouse central nervous system. The Journal of Neuroscience 1993; 13:3006-17.
180. Firmbach Kraft I, Byers M, Shows T et al. tyk2, prototype of a novel class of non-receptor tyrosine kinase genes. Oncogene 1990; 5:1329-36.
181. Lai KS, Jin Y, Graham DK et al. A kinase-deficient splice variant of the human JAK3 is expressed in hematopoietic and epithelial cancer cells. J Biol Chem 1995; 270:25028-36.
182. Sanchez MP, Tapley P, Saini SS et al. Multiple tyrosine protein kinases in rat hippocampal neurons: isolation of Ptk-3, a receptor expressed in proliferative zones of the developing brain. Proc Natl Acad Sci USA 1994; 91:1819-23.
183. Pritchard MA, Baker E, Callen DF et al. Two members of the JAK family of protein tyrosine kinases map to chromosomes 1p31.3 and 9p24. Mamm Genome 1992; 3:36-8.
184. Gough NM, Rakar S, Harpur A et al. Localization of genes for two members of the JAK family of protein tyrosine kinases to murine chromosomes 4 and 19. Mamm Genome 1995; 6:247-8.
185. Sadowski HB, Stone JC, Pawson T. A noncatalytic domain conserved among cytoplasmic protein-tyrosine kinases modifies the kinase function and transforming activity of Fujinami sarcoma virus $P130^{gag\text{-}fps}$. Mol Cell Biol 1986; 6:4396-408.
186. Kozma SC, Redmond SMS, Xiao-Chang F et al. Activation of the receptor kinase domain of the *trk* oncogene by recombination with two different cellular sequences. EMBO 1988; 7:147-54.
187. Knighton DR, Zheng JH, Ten Eyck LF et al. Crystal structure of the catalytic subunit of cyclic adenosine monophosphate-dependent protein kinase [see comments]. Science 1991; 253:407-14.
188. Sternberg MJE, Taylor WR. Modelling the ATP-binding site of oncogene products, the epidermal growth factor receptor and related proteins. FEBS Lett 1984; 175:387-92.
189. Verderame MF, Varmus HE. Highly conserved amino acids in the SH2 and catalytic domains of v-src are altered in naturally occurring, transformation-defective alleles. Oncogene 1994; 9:175-82.
190. Hubbard SR, Wei L, Ellis L et al. Crystal structure of the tyrosine kinase domain of the human insulin receptor. Nature 1994; 372:746-54.

191. Chinkers M, Garbers DL. The protein kinase domain of the ANP receptor is required for signaling. Science 1989; 245:1392-4.
192. Bernards A. Predicted *tyk2* protein contains two tandem protein kinases domains. Oncogene 1991; 6:1185-7.
193. Jones SW, Erikson E, Blenis J et al. A *Xenopus* ribosomal protein S6 kinase has two apparent kinase domains that are each similar to distinct protein kinases. Proc Natl Acad Sci USA 1988; 85:3377-81.
194. Wek RC, Jackson BM, Hinnebusch AG. Juxtaposition of domains homologous to protein kinases and histidyl-tRNA synthetases in GCN2 protein suggests a mechanism for coupling *GCN4* expression to amino acid availability. Proc Natl Acad Sci USA 1989; 86:4579-83.
195. Kraus MH, Issing W, Miki T et al. Isolation and characterization of *ERBB3*, a third member of the *ERBB*/epidermal growth factor receptor family: evidence for overexpression in a subset of human mammary tumors. Proc Natl Acad Sci USA 1989; 86:9193-7.
196. Moran MF, Koch CA, Sadowski I et al. Mutational analysis of a phosphotransfer motif essential for v-fps tyrosine kinase activity. Oncogene 1988; 3:665-72.
197. Ruff Jamison S, Chen K, Cohen S. Epidermal growth factor induces the tyrosine phosphorylation and nuclear translocation of Stat 5 in mouse liver. Proc Natl Acad Sci USA 1995; 92:4215-8.
198. Zhao Y, Wagner F, Frank SA et al. The amino-terminal portion of the JAK2 protein kinase is necessary for binding and phosphorylation of the granulocyte-macrophage colony stimulating factor β chain. J Biol Chem 1995; 270:13814-8.
199. Velazquez L, Mogensen KE, Barbieri G et al. Distinct domains of the protein tyrosine kinase tyk2 required for binding of interferon-alpha/beta and for signal transduction. J Biol Chem 1995; 270:3327-34.
200. Brugge JS, Erikson RL. Identification of a transformation-specific antigen induced by an avian sarcoma virus. Nature 1977; 269:346-8.
201. Pelicci G, Lanfrancone L, Grignani F et al. A novel transforming protein (SHC) with an SH2 domain is implicated in mitogenic signal transduction. Cell 1992; 70:93-104.
202. Schindler C, Fu X-F, Improta T et al. Proteins of transcription factor ISGF-3: One gene encodes the 91- and 84-kDa ISGF-3 proteins that are activated by interferon α. Proc Natl Acad Sci USA 1992; 89:7836-9.
203. Fu X-Y, Kessler DS, Veals SA et al. ISGF3, the transcriptional activator induced by interferon α, consists of multiple interacting polypeptide chains. Proc Natl Acad Sci USA 1992; 87:8555-9.
204. Zhong Z, Wen Z, Darnell JEJ. Stat3 and Stat4: Members of the family of signal transducers and activators of transcription. Proc Natl Acad Sci USA 1994; 91:4806-10.

205. Zhong Z, Wen Z, Darnell JEJ. Stat3:A STAT family member activated by tyrosine phosphorylation in response to epidermal growth factor and interleukin-6. Science 1994; 264:95-8.
206. Akira S, Nishio Y, Inoue M et al. Molecular cloning of APRF, a novel IFN-stimulated gene factor 3 p91-related transcription factor involved in the gp130-mediated signaling pathway. Cell 1994; 77:63-71.
207. Wakao H, Gouilleux F, Groner B. Mammary gland factor (MGF) is a novel member of the cytokine regulated transcription factor gene family and confers the prolactin response. EMBO J 1994; 13:2182-91.
208. Hou J, Schindler U, Henzel WJ et al. An interleukin-4-induced transcription factor: IL-4 Stat. Science 1994; 265:1701-6.
209. Levy DE, Kessler DS, Pine R et al. Cytoplasmic activation of ISGF3, the positive regulator of interferon alpha-stimulated transcription, reconstituted in vitro. Genes Dev 1989; 3:1362-71.
210. Dale TC, Imam AM, Kerr IM et al. Rapid activation by interferon alpha of a latent DNA-binding protein present in the cytoplasm of untreated cells. Proc Natl Acad Sci USA 1989; 86:1203-7.
211. Fu X-Y, Schindler C, Improta T et al. The proteins of ISGF-3, the interferon α-induced transcriptional activator, define a gene family involved in signal transduction. Proc Natl Acad Sci USA 1992; 89:7840-3.
212. Larner AC, Jonak G, Cheng YS et al. Transcriptional induction of two genes in human cells by beta interferon. Proc Natl Acad Sci USA 1984; 81:6733-7.
213. Larner AC, Chaudhuri A, Darnell JE. Transcriptional induction by interferon. New protein(s) determine the extent and length of the induction. J Biol Chem 1986; 261:453-9.
214. Decker T, Lew DJ, Cheng YS et al. Interactions of alpha- and gamma-interferon in the transcriptional regulation of the gene encoding a guanylate-binding protein. EMBO J 1989; 8:2009-14.
215. Reid LE, Brasnett AH, Gilbert CS et al. A single DNA response element can confer inducibility by both α- and γ-interferons. Proc Natl Acad Sci USA 1989; 86:840-4.
216. Lew DJ, Decker T, Strehlow I et al. Overlapping elements in the Guanylate-binding Protein gene promoter mediate transcriptional induction by alpha and gamma interferons.. Mol Cell Biol 1991; 11:182-91.
217. Shuai K, Schindler C, Prezioso VR et al. Activation of transcription by IFN-γ: tyrosine phosphorylation of a 91-kDa DNA binding protein. Science 1992; 258:1808-12.
218. Fu X-Y. A transcription factor with SH2 and SH3 domains is directly activated by an interferon α-induced cytoplasmic protein tyrosine kinase(s). Cell 1992; 70:323-35.

219. Gutch MJ, Daly C, Reich NC. Tyrosine phosphorylation is required for activation of an α interferon-stimulated transcription factor. Proc Natl Acad Sci USA 1992; 89:11411-5.
220. Veals SA, Schindler C, Leonard D et al. Subunit of an alpha-interferon-responsive transcription factor is related to interferon regulatory factor and Myb families of DNA-binding proteins. Mol Cell Biol 1992; 12:3315-24.
221. Darnell JE Jr, Kerr IM, Stark GR. Jak-STAT pathways and transcriptional activation in response to IFNs and other extracellular signaling proteins. Science 1994; 264:1415-21.
222. Lutticken C, Wegenka UM, Yuan J et al. Association of transcription factor APRF and protein kinase Jak1 with the interleukin-6 signal transducer gp130. Science 1994; 263:89-92.
223. Schaefer TS, Sanders L, Nathans D. Cooperative transcriptional activity of Jun and Stat3b, a short form of Stat3. Proc Natl Acad Sci USA 1995; 92:9097-101.
224. Yamamoto K, Quelle FW, Thierfelder WE et al. Stat4, a novel gamma interferon activation site-binding protein expressed in early myeloid differentiation. Mol Cell Biol 1994; 14:4342-9.
225. Gilmour KC, Reich NC. Receptor to nucleus signaling by prolactin and interleukin 2 via activation of latent DNA-binding factors. Proc Natl Acad Sci USA 1994; 91:6850-4.
226. David M, Petricoin EF, Igarashi K-I et al. Prolactin activates the interferon-regulated p91 transcription factor and the Jak2 kinase by tyrosine phosphorylation. Proc Natl Acad Sci USA 1994; 91:7174-8.
227. Gouilleux F, Wakao H, Mundt M et al. Prolactin induces phosphorylation of Tyr694 of Stat5 (MGF), a prerequisite for DNA binding and induction of transcription. EMBO J 1994; 13:4361-9.
228. Mui AL, Wakao H, O'Farrell AM et al. Interleukin-3, granulocyte-macrophage colony stimulating factor and interleukin-5 transduce signals through two STAT5 homologs. EMBO J 1995; 14:1166-75.
229. Mui AL, Wakao H, Harada N et al. Interleukin-3, granulocyte-macrophage colony-stimulating factor, and interleukin-5 transduce signals through two forms of STAT5. J Leukoc Biol 1995; 57:799-803.
230. Schindler C, Kashleva H, Pernis A et al. STF-IL-4: a novel IL-4-induced signal transducing factor. EMBO 1994; 13:1350-6.
231. Quelle FW, Shimoda K, Thierfelder W et al. Cloning of murine Stat6 and human Stat6, Stat proteins that are tyrosine phosphorylated in responses to IL-4 and IL-3 but are not required for mitogenesis. Mol Cell Biol 1995; 15:3336-43.

232. Adams MD, Kelley JM, Gocayne JD et al. Complementary DNA sequencing: expressed sequence tags and human genome project. Science 1991; 252:1651-6.
233. Adams MD, Kerlavage AR, Fleischmann RD et al. Initial assessment of human gene diversity and expression patterns based upon 83 million nucleotides of cDNA sequence. Nature 1995; 377:3-174.
234. Adams MD, Dubnick M, Kerlavage AR et al. Sequence identification of 2,375 human brain genes [see comments]. Nature 1992; 355:632-4.
235. Adams MD, Kerlavage AR, Fields C et al. 3,400 new expressed sequence tags identify diversity of transcripts in human brain. Nat Genet 1993; 4:256-67.
236. Yan R, Qureshi S, Zhong Z et al. The genomic structure of the STAT genes: multiple exons in coincident sites in Stat1 and Stat2. Nucleic Acids Res 1995; 23:459-63.
237. Improta T, Schindler C, Horvath CM et al. Transcription factor ISGF-3 formation requires phosphorylated Stat91 protein, but Stat113 protein is phosphorylated independently of Stat91 protein. Proc Natl Acad Sci USA 1994; 91:4776-80.
238. Shuai K, Horvath CM, Tsai Huang LH et al. Interferon activation of the transcription factor Stat91 involves dimerization through SH2-phosphotyrosyl peptide interactions. Cell 1994; 76:821-8.
239. Heim MH, Kerr IM, Stark GR et al. Contribution of STAT SH2 groups to specific interferon signaling by the Jak-STAT pathway. Science 1995; 267:1347-9.
240. Shuai K, Stark GR, Kerr IM et al. A single phosphotyrosine residue of Stat91 is required for gene activation by interferon-γ. Science 1993; 261:1744-6.
241. Wen Z, Zhong Z, Darnell JEJ. Maximal activation of transcription by Stat1 and Stat3 requires both tyrosine and serine phosphorylation. Cell 1995; 82:241-50.
242. Horvath CM, Wen Z, Darnell JE. A STAT protein domain that determines DNA sequence recognition suggests a novel DNA-binding domain. Genes Dev 1995; 9:984-94.
243. Tani T, Fujii S, Inoue S et al. Binding mode of phospholipase A2 with a new type of phospholipid analog having an oxazolidinone ring. J Biochem Tokyo 1995; 117:176-82.
244. Yu H, Rosen MK, Shin TB et al. Solution structure of the SH3 domain of Src and identification of its ligand-binding site. Science 1992; 258:1665-8.
245. Yu H, Chen JK, Feng S et al. Structural basis for the binding of proline-rich peptides to SH3 domains. Cell 1994; 76:933-45.
246. Greenlund AC, Farrar MA, Viviano BL et al. Ligand-induced IFNγ receptor tyrosine phosphorylation couples the receptor to its signal transduction system (p91). EMBO 1994; 13:1591-600.

247. David M, Romero G, Zhang ZY et al. In vitro activation of the transcription factor ISGF3 by interferon alpha involves a membrane-associated tyrosine phosphatase and tyrosine kinase. J Biol Chem 1993; 268:6593-9.
248. Larner AC, David M, Feldman GM et al. Tyrosine phosphorylation of DNA binding proteins by multiple cytokines. Science 1993; 261:1730-3.
249. Eilers A, Georgellis D, Klose B et al. Differentiation-regulated serine phosphorylation of STAT1 promotes GAF activation in macrophages. Mol Cell Biol 1995; 15:3579-86.
250. Shuai K, Ziemiecki A, Wilks AF et al. Polypeptide signaling to the nucleus through tyrosine phosphorylation of Jak and Stat proteins. Nature 1993; 366:580-3.
251. Shuai K, Stark GR, Kerr IM et al. A single phosphotyrosine residue of Stat91 required for gene activation by interferon-gamma [see comments]. Science 1993; 261:1744-6.
252. Jacobson NG, Szabo SJ, Weber Nordt RM et al. Interleukin 12 signaling in T helper type 1 (Th1) cells involves tyrosine phosphorylation of signal transducer and activator of transcription (Stat)3 and Stat4. J Exp Med 1995; 181:1755-62.
253. Bacon CM, Petricoin EF, Ortaldo JR et al. Interleukin 12 induces tyrosine phosphorylation of STAT4 in human lymphocytes. Proc Natl Acad Sci USA 1995; 92:7307-11.
254. Zhang X, Blenis J, Li HC et al. Requirement of serine phosphorylation for formation of STAT-promoter complexes. Science 1995; 267:1990-4.
255. Clark Lewis I, Sanghera JS, Pelech SL. Definition of a consensus sequence for peptide substrate recognition by p44mapk, the meiosis-activated myelin basic protein kinase. J Biol Chem 1991; 266:15180-4.
256. Sadowski HB, Shuai K, Darnell JE, Jr. et al. A common nuclear signal transduction pathway activated by growth factor and cytokine receptors [see comments]. Science 1993; 261:1739-44.
257. Rothman P, Kreider B, Azam M et al. Cytokines and growth factors signal through tyrosine phosphorylation of a family of related transcription factors. Immunity 1994; 1:457-68.
258. Wagner BJ, Hayes TE, Hoban CJ et al. The SIF binding element confers sis/PDGF inducibility onto the c-fos promoter. EMBO J 1990; 9:4477-84.
259. Schindler C, Darnell JE, Jr. Transcriptional responses to polypeptide ligands: The JAK- STAT pathway. Annual Review of Biochemistry 1995; 64:621-51.
260. Pestka S, Langer JA, Zoon KC et al. Interferons and their actions. Ann Rev Biochem 1987; 56:727-77.

261. Reich NC, Darnell JE. Differential binding of interferon-induced factors to an oligonucleotide that mediates transcriptional activation. Nucleic Acids Res 1989; 17:3415-24.
262. Dale TC, Rosen JM, Guille MJ et al. Overlapping sites for constitutive and induced DNA binding factors involved in interferon-stimulated transcription. EMBO 1989; 8:831-9.
263. Seidel HM, Milocco LH, Lamb P et al. Spacing of palindromic half sites as a determinant of selective STAT (signal transducers and activators of transcription) DNA binding and transcriptional activity. Proc Natl Acad Sci USA 1995; 92:3041-5.
264. Schindler U, Wu P, Rothe M et al. Components of a Stat recognition code: evidence for two layers of molecular selectivity. Immunity 1995; 2:689-97.
265. Qureshi SA, Salditt Georgieff M, Darnell JE. Tyrosine-phosphorylated Stat1 and Stat2 plus a 48-kDa protein all contact DNA in forming interferon-stimulated-gene factor 3. Proc Natl Acad Sci USA 1995; 92:3829-33.
266. Wakao H, Harada N, Kitamura T et al. Interleukin 2 and erythropoietin activate STAT5/MGF via distinct pathways. EMBO J 1995; 14:2527-35.
267. Pallard C, Gouilleux F, Charon M et al. Interleukin-3, erythropoietin, and prolactin activate a STAT5-like factor in lymphoid cells. J Biol Chem 1995; 270:15942-5.
268. Gouilleux F, Pallard C, Dusanter Fourt I et al. Prolactin, growth hormone, erythropoietin and granulocyte-macrophage colony stimulating factor induce MGF-Stat5 DNA binding activity. EMBO J 1995; 14:2005-13.
269. Gurney AL, Wong SC, Henzel WJ et al. Distinct regions of c-Mpl cytoplasmic domain are coupled to the JAK-STAT signal transduction pathway and Shc phosphorylation. Proc Natl Acad Sci USA 1995; 92:5292-6.
270. Pallard C, Gouilleux F, Benit L et al. Thrombopoietin activates a STAT5-like factor in hematopoietic cells. EMBO J 1995; 14:2847-56.
271. Kowalski CJ, Schneiderman ED, Willis SM. PC program implementing an alternative to the paired t-test which adjusts for regression to the mean. Int J Biomed Comput 1994; 37:189-94.
272. Burdon TG, Demmer J, Clark AJ et al. The mammary factor MPBF is a prolactin-induced transcriptional regulator which binds to STAT factor recognition sites. FEBS Lett 1994; 350:177-82.
273. Wegenka UM, Buschmann J, Lutticken C et al. Acute-phase response factor, a nuclear factor binding to acute-phase response elements, is rapidly activated by interleukin-6 at the posttranslational level. Mol Cell Biol 1993; 13:276-88.

274. Hou J, Schindler U, Henzel WJ et al. Identification and purification of human Stat proteins activated in response to interleukin-2. Immunity 1995; 2:321-9.
275. Yin T, Yang L, Yang YC. Tyrosine phosphorylation and activation of JAK family tyrosine kinases by interleukin-9 in MO7E cells. Blood 1995; 85:3101-6.
276. Ihle JN, Witthuhn BA, Quelle FW et al. Signaling through the hematopoietic cytokine receptors. Annu Rev Immunol 1995; 13:369-98.
277. Sadowski HB, Gilman MZ. Cell-free activation of a DNA-binding protein by epidermal growth factor. Nature 1993; 362:79-83.
278. Ruff Jamison S, Zhong Z, Wen Z et al. Epidermal growth factor and lipopolysaccharide activate Stat3 transcription factor in mouse liver. J Biol Chem 1994; 269:21933-5.
279. Silvennoinen O, Schindler C, Schlessinger J et al. Ras-independent growth factor signaling by transcription factor tyrosine phosphorylation [see comments]. Science 1993; 261:1736-9.
280. Ruff-Jamison S, Chen K, Cohen S. Induction by EGF and interferon-γ of tyrosine phosphorylated DNA binding proteins in mouse liver nuclei. Science 1993; 261:1733-6.
281. Shou C, Wurmser A, Ling K et al. Differential response of the Ras exchange factor, Ras-GRF to tyrosine kinase and G protein mediated signals. Oncogene 1995; 10:1887-93.
282. Schindler C, Shuai K, Prezioso VR et al. Interferon-dependent tyrosine phosphorylation of a latent cytoplasmic transcription factor. Science 1992; 257:809-13.
283. Decker T, Lew DJ, Mirkovitch J et al. Cytoplasmic activation of GAF, an IFN-gamma-regulated DNA-binding factor. EMBO J 1991; 10:927-32.
284. Leung S, Qureshi SA, Kerr IM et al. Role of STAT2 in the alpha interferon signaling pathway. Mol Cell Biol 1995; 15:1312-7.
285. Muller M, Laxton C, Briscoe J et al. Complementation of a mutant cell line: central role of the 91 kDa polypeptide of ISGF3 in the interferon-α and -γ signal transduction pathways. EMBO J 1993; 12:4221-8.
286. Hayes TE, Kitchen AM, Cochran BH. Inducible binding of a factor to the c-fos regulatory region. Proc Natl Acad Sci USA 1987; 84:1272-6.
287. Silvennoinen O, Schindler C, Schlessinger J et al. Ras-independent growth factor signaling by transcription factor tyrosine phosphorylation. Science 1993; 261:1736-9.
288. Fu X-Y, Zhang J. Transcription factor p91 interacts with the epidermal growth factor receptor and mediates activation of the *c-fos* gene promoter. Cell 1993; 74:1135-45.

289. Pellegrini S, John J, Shearer M et al. Use of a selectable marker regulated by alpha interferon to obtain mutations in the signaling pathway. Mol Cell Biol 1989; 9:4605-12.
290. John J, McKendry R, Pellegrini S et al. Isolation and characterization of a new mutant human cell line unresponsive to alpha and beta interferons. Mol Cell Biol 1991; 11:4189-95.
291. McKendry R, John J, Flavell D et al. High-frequency mutagenesis of human cells and characterization of a mutant unresponsive to both alpha and gamma interferons. Proc Natl Acad Sci USA 1991; 88:11455-9.
292. Porter ACG, Chernajovsky Y, Dale TC et al. Interferon response element of the human gene 6-16. EMBO 1988; 7:85-92.
293. Ackrill AM, Reid LE, Gilbert CS et al. Differential response of the human 6-16 and 9-27 genes to alpha and gamma interferons. Nucleic Acids Res 1991; 19:591-8.
294. Mao C, Davies D, Kerr IM et al. Mutant human cells defective in induction of major histocompatibility complex class II genes by interferon gamma. Proc Natl Acad Sci USA 1993; 90:2880-4.
295. Watling D, Guschin D, Muller M et al. Complementation by the protein tyrosine kinase JAK2 of a mutant cell line defective in the interferon-γ signal transduction pathway. Nature 1993; 366:166-70.
296. Loh JE, Schindler C, Ziemiecki A et al. Mutant cell lines unresponsive to alpha/beta and gamma interferon are defective in tyrosine phosphorylation of ISGF-3 alpha components. Mol Cell Biol 1994; 14:2170-9.
297. Maclennan AJ, Shaw G. A yeast SH2 domain. TIBS 1993; 18:464-5.
298. Velazquez L, Fellous M, Stark GR et al. A protein tyrosine kinase in the interferon α/β signaling pathway. Cell 1992; 70:313-22.
299. Velazquez L, Fellous M, Stark GR et al. A protein tyrosine kinase in the interferon alpha/beta signaling pathway. Cell 1992; 70:313-22.
300. Muller M, Briscoe J, Laxton C et al. The protein tyrosine kinase JAK1 complements defects in interferon-alpha/beta and -gamma signal transduction [see comments]. Nature 1993; 366:129-35.
301. Shual K, Ziemiecki A, Wilks AF et al. Polypeptide signaling to the nucleus through tyrosine phosphorylation of Jak and Stat proteins. Nature 1993; 366:580-3.
302. Beadling C, Guschin D, Witthuhn BA et al. Activation of JAK kinases and STAT proteins by interleukin-2 and interferon alpha, but not the T cell antigen receptor, in human T lymphocytes. EMBO J 1994; 13:5605-15.
303. Colamonici O, Yan H, Domanski P et al. Direct binding to and tyrosine phosphorylation of the alpha subunit of the type I inter-

feron receptor by p135tyk2 tyrosine kinase. Mol Cell Biol 1994; 14:8133-42.
304. Silva CM, Lu H, Weber MJ et al. Differential tyrosine phosphorylation of JAK1, JAK2, and STAT1 by growth hormone and interferon-gamma in IM-9 cells. J Biol Chem 1994; 269:27532-9.
305. Jahnke A, Johnson JP. Synergistic activation of intercellular adhesion molecule 1 (ICAM-1) by TNF-alpha and IFN-gamma is mediated by p65/p50 and p65/c-Rel and interferon-responsive factor Stat1 alpha (p91) that can be activated by both IFN-gamma and IFN-alpha. FEBS Lett 1994; 354:220-6.
306. Raz R, Durbin JE, Levy DE. Acute phase response factor and additional members of the interferon-stimulated gene factor 3 family integrate diverse signals from cytokines, interferons, and growth factors. J Biol Chem 1994; 269:24391-5.
307. Colamonici OR, Platanias LC, Domanski P et al. Transmembrane signaling by the alpha subunit of the type I interferon receptor is essential for activation of the JAK kinases and the transcriptional factor ISGF3. J Biol Chem 1995; 270:8188-93.
308. Barbieri G, Velazquez L, Scrobogna M et al. Activation of the protein tyrosine kinase tyk2 by interferon alpha/beta. Eur J Biochem 1994; 223:427-35.
309. Shuai K, Ziemiecki A, Wilks AF et al. Polypeptide signaling to the nucleus through tyrosine phosphorylation of Jak and Stat proteins. Nature 1993; 366:580-3.
310. Yuan J, Wegenka UM, Lutticken C et al. The signaling pathways of interleukin-6 and gamma interferon converge by the activation of different transcription factors which bind to common responsive DNA elements. Mol Cell Biol 1994; 14:1657-68.
311. Tourkine N, Schindler C, Larose M et al. Activation of STAT factors by prolactin, interferon-gamma, growth hormones, and a tyrosine phosphatase inhibitor in rabbit primary mammary epithelial cells. J Biol Chem 1995; 270:20952-61.
312. Barahmand pour F, Meinke A, Eilers A et al. Colony-stimulating factors and interferon-gamma activate a protein related to MGF-Stat 5 to cause formation of the differentiation-induced factor in myeloid cells. FEBS Lett 1995; 360:29-33.
313. Sakatsume M, Igarashi K, Winestock KD et al. The Jak kinases differentially associate with the alpha and beta (accessory factor) chains of the interferon gamma receptor to form a functional receptor unit capable of activating STAT transcription factors. J Biol Chem 1995; 270:17528-34.
314. Greenlund AC, Morales MO, Viviano BL et al. Stat recruitment by tyrosine-phosphorylated cytokine receptors: an ordered reversible affinity-driven process. Immunity 1995; 2:677-87.

315. Novick D, Cohen B, Rubinstein M. The human interferon alpha/beta receptor: characterization and molecular cloning. Cell 1994; 77:391-400.
316. Witthuhn BA, Quelle FW, Silvennoinen O et al. JAK2 associates with the erythropoietin receptor and is tyrosine phosphorylated and activated following stimulation with erythropoietin. Cell 1993; 74:227-36.
317. Miura O, Nakamura N, Quelle FW et al. Erythropoietin induces association of the JAK2 protein tyrosine kinase with the erythropoietin receptor in vivo. Blood 1994; 84:1501-7.
318. Barber DL, D'Andrea AD. Erythropoietin and interleukin-2 activate distinct JAK kinase family members. Mol Cell Biol 1994; 14:6506-14.
319. Ohashi T, Masuda M, Ruscetti SK. Induction of sequence-specific DNA-binding factors by erythropoietin and the spleen focus-forming virus. Blood 1995; 85:1454-62.
320. Nicholson SE, Oates AC, Harpur AG et al. Tyrosine kinase JAK1 is associated with the granulocyte-colony-stimulating factor receptor and both become tyrosine-phosphorylated after receptor activation. Proc Natl Acad Sci USA 1994; 91:2985-8.
321. Matsuda T, Hirano T. Association of p72 tyrosine kinase with Stat factors and its activation by interleukin-3, interleukin-6, and granulocyte colony-stimulating factor. Blood 1994; 83:3457-61.
322. Tian SS, Lamb P, Seidel HM et al. Rapid activation of the STAT3 transcription factor by granulocyte colony-stimulating factor. Blood 1994; 84:1760-4.
323. Shimoda K, Iwasaki H, Okamura S et al. G-CSF induces tyrosine phosphorylation of the JAK2 protein in the human myeloid G-CSF responsive and proliferative cells, but not in mature neutrophils. Biochem Biophys Res Commun 1994; 203:922-8.
324. Argetsinger LS, Campbell GS, Yang X et al. Identification of JAK2 as a growth hormone receptor-associated tyrosine kinase. Cell 1993; 74:237-44.
325. Hackett RH, Wang YD, Larner AC. Mapping of the cytoplasmic domain of the human growth hormone receptor required for the activation of Jak2 and stat proteins. J Biol Chem 1995; 270: 21326-30.
326. Wang Y, Morella KK, Ripperger J et al. Receptors for interleukin-3 (IL-3) and growth hormone mediate an IL-6-type transcriptional induction in the presence of JAK2 or STAT3. Blood 1995; 86:1671-9.
327. Campbell GS, Argetsinger LS, Ihle JN et al. Activation of JAK2 tyrosine kinase by prolactin receptors in Nb2 cells and mouse mammary gland explants. Proc Natl Acad Sci USA 1994; 91:5232-6.

328. Rui H, Kirken RA, Farrar WL. Activation of receptor-associated tyrosine kinase JAK2 by prolactin. J Biol Chem 1994; 269:5364-8.
329. DaSilva L, Howard OMZ, Rui H et al. Growth signaling and JAK2 association mediated by membrane-proximal cytoplasmic regions of prolactin receptors. J Biol Chem 1994; 269:18267-70.
330. Dusanter Fourt I, Muller O, Ziemiecki A et al. Identification of JAK protein tyrosine kinases as signaling molecules for prolactin. Functional analysis of prolactin receptor and prolactin-erythropoietin receptor chimera expressed in lymphoid cells. EMBO J 1994; 13:2583-91.
331. Bacon CM, Tortolani PJ, Shimosaka A et al. Thrombopoietin (TPO) induces tyrosine phosphorylation and activation of STAT5 and STAT3. FEBS Lett 1995; 370:63-8.
332. Sattler M, Durstin MA, Frank DA et al. The thrombopoietin receptor c-MPL activates JAK2 and TYK2 tyrosine kinases. Experimental Hematology 1995; 23:1040-8.
333. Morella KK, Bruno E, Kumaki S et al. Signal transduction by the receptors for thrombopoietin (c-mpL) and interleukin-3 in hematopoietic and nonhematopoietic cells. Blood 1995; 86:557-71.
334. Stahl N, Boulton TG, Farruggella T et al. Association and activation of Jak-Tyk kinases by CNTF-LIF-IL-6 β receptor components. Science 1994; 263:92-5.
335. Bonni A, Frank DA, Schindler C et al. Characterization of a pathway for ciliary neurotrophic factor signaling to the nucleus. Science 1993; 262:1575-9.
336. Hirano T, Matsuda T, Nakajima K. Signal transduction through gp130 that is shared among the receptors for the interleukin 6 related cytokine subfamily. Stem Cells Dayt 1994; 12:262-77.
337. Coffer P, Lutticken C, van Puijenbroek A et al. Transcriptional regulation of the junB promoter: analysis of STAT-mediated signal transduction. Oncogene 1995; 10:985-94.
338. Symes AJ, Rajan P, Corpus L et al. C/EBP-related sites in addition to a STAT site are necessary for ciliary neurotrophic factor-leukemia inhibitory factor-dependent transcriptional activation by the vasoactive intestinal peptide cytokine response element. J Biol Chem 1995; 270:8068-75.
339. Symes A, Lewis S, Corpus L et al. STAT proteins participate in the regulation of the vasoactive intestinal peptide gene by the ciliary neurotrophic factor family of cytokines. Mol Endocrinol 1994; 8:1750-63.
340. Murakami M, Hibi M, Nakagawa N et al. IL-6-induced homodimerization of gp130 and associated activation of a tyrosine kinase. Science 1993; 260:1808-10.

341. Standke GJ, Meier VS, Groner B. Mammary gland factor activated by prolactin on mammary epithelial cells and acute-phase response factor activated by interleukin-6 in liver cells share DNA. Mol Endocrinol 1994; 8:469-77.
342. Fujitani Y, Nakajima K, Kojima H et al. Transcriptional activation of the IL-6 response element in the junB promoter is mediated by multiple Stat family proteins. Biochem Biophys Res Commun 1994; 202:1181-7.
343. Wegenka UM, Lutticken C, Buschmann J et al. The interleukin-6-activated acute-phase response factor is antigenically and functionally related to members of the signal transducer and activator of transcription (STAT) family. Mol Cell Biol 1994; 14:3186-96.
344. Harroch S, Gothelf Y, Revel M et al. 5' upstream sequences of MyD88, an IL-6 primary response gene in M1 cells: Detection of functional IRF-1 and Stat factors binding sites. Nucl Acid Res 1995; 23:3539-46.
345. Nakajima K, Matsuda T, Fujitani Y et al. Signal transduction through IL-6 receptor: involvement of multiple protein kinases, stat factors, and a novel H7-sensitive pathway. Ann N Y Acad Sci 1995; 762:55-70.
346. Henttinen T, Levy DE, Silvennoinen O et al. Activation of the signal transducer and transcription (STAT) signaling pathway in a primary T cell response. Critical role for IL-6. J Immunol 1995; 155:4582-7.
347. Harroch S, Revel M, Chebath J. Interleukin-6 signaling via four transcription factors binding palindromic enhancers of different genes. J Biol Chem 1994; 269:26191-5.
348. Berger LC, Hawley TS, Lust JA et al. Tyrosine phosphorylation of JAK-TYK kinases in malignant plasma cell lines growth-stimulated by interleukins 6 and 11. Biochem Biophys Res Commun 1994; 202:596-605.
349. Fuhrer DK, Feng GS, Yang YC. Syp associates with gp130 and Janus kinase 2 in response to interleukin-11 in 3T3-L1 mouse preadipocytes. J Biol Chem 1995; 270:24826-30.
350. Boulton TG, Stahl N, Yancopoulos GD. Ciliary neurotrophic factor/leukemia inhibitory factor/interleukin 6/oncostatin M family of cytokines induces tyrosine phosphorylation of a common set of proteins overlapping those induced by other cytokines and growth factors. J Biol Chem 1994; 269:11648-55.
351. Silvennoinen O, Witthuhn BA, Quelle FW et al. Structure of the murine Jak2 protein-tyrosine kinase and its role in interleukin 3 signal transduction. Proc Natl Acad Sci USA 1993; 90:8429-33.
352. van der Bruggen T, Caldenhoven E, Kanters D et al. Interleukin-5 signaling in human eosinophils involves JAK2 tyrosine kinase and Stat1 alpha. Blood 1995; 85:1442-8.

353. Kirken RA, Rui H, Malabarba MG et al. Identification of interleukin-2 receptor-associated tyrosine kinase p116 as novel leukocyte-specific Janus kinase. J Biol Chem 1994; 269:19136-41.
354. Nielsen M, Svejgaard A, Skov S et al. Interleukin-2 induces tyrosine phosphorylation and nuclear translocation of stat3 in human T lymphocytes. Eur J Immunol 1994; 24:3082-6.
355. Musso T, Johnston JA, Linnekin D et al. Regulation of JAK3 expression in human monocytes: phosphorylation in response to interleukins 2, 4, and 7. J Exp Med 1995; 181:1425-31.
356. Witthuhn BA, Silvennoinen O, Miura O et al. Involvement of the Jak-3 Janus kinase in signaling by interleukins 2 and 4 in lymphoid and myeloid cells. Nature 1994; 370:153-7.
357. Brunn GJ, Falls EL, Nilson AE et al. Protein-tyrosine kinase-dependent activation of STAT transcription factors in interleukin-2- or interleukin-4-stimulated T lymphocytes. J Biol Chem 1995; 270:11628-35.
358. Pernis A, Gupta S, Yopp J et al. Gamma chain-associated cytokine receptors signal through distinct transducing factors. J Biol Chem 1995; 270:14517-22.
359. Malabarba MG, Kirken RA, Rui H et al. Activation of JAK3, but not JAK1, is critical to interleukin-4 (IL4) stimulated proliferation and requires a membrane-proximal region of IL4 receptor alpha. J Biol Chem 1995; 270:9630-7.
360. Rolling C, Treton D, Beckmann P et al. JAK3 associates with the human interleukin 4 receptor and is tyrosine phosphorylated following receptor triggering. Oncogene 1995; 10:1757-61.
361. Yin T, Tsang ML, Yang YC. JAK1 kinase forms complexes with interleukin-4 receptor and 4PS/insulin receptor substrate-1-like protein and is activated by interleukin-4 and interleukin-9 in T lymphocytes. J Biol Chem 1994; 269:26614-7.
362. Fenghao X, Saxon A, Nguyen A et al. Interleukin 4 activates a signal transducer and activator of transcription (Stat) protein which interacts with an interferon-gamma activation site-like sequence upstream of the I epsilon exon in a human B cell line. Evidence for the involvement of Janus kinase 3 and interleukin-4 Stat. J Clin Invest 1995; 96:907-14.
363. Schindler U, Wu P, Rothe M et al. Components of a Stat recognition code: evidence for two layers of molecular selectivity. Immunity 1995; 2:689-97.
364. Berton MT, Linehan LA. IL-4 activates a latent DNA-binding factor that binds a shared IFN-gamma and IL-4 response element present in the germ-line gamma 1 Ig promoter. J Immunol 1995; 154:4513-25.

365. Welham MJ, Learmonth L, Bone H et al. Interleukin-13 signal transduction in lymphohemopoietic cells. Similarities and differences in signal transduction with interleukin-4 and insulin. J Biol Chem 1995; 270:12286-96.
366. Foxwell BMJ, Beadling C, Guschin D et al. Interleukin-7 can induce the activation of Jak 1, Jak 3 and STAT 5 proteins in murine T cells. European Journal of Immunology 1995; 25:3041-6.
367. Lin JX, Migone TS, Tsang M et al. The role of shared receptor motifs and common Stat proteins in the generation of cytokine pleiotropy and redundancy by IL-2, IL-4, IL-7, IL-13, and IL-15. Immunity 1995; 2:331-9.
368. Zeng YX, Takahashi H, Shibata M et al. JAK3 Janus kinase is involved in interleukin 7 signal pathway. FEBS Lett 1994; 353:289-93.
369. Yin T, Keller SR, Quelle FW et al. Interleukin-9 induces tyrosine phosphorylation of insulin receptor substrate-1 via JAK tyrosine kinases. J Biol Chem 1995; 270:20497-502.
370. Bacon CM, McVicar DW, Ortaldo JR et al. Interleukin 12 (IL-12) induces tyrosine phosphorylation of JAK2 and TYK2: differential use of Janus family tyrosine kinases by IL-2 and IL-12. J Exp Med 1995; 181:399-404.
371. Smerz Bertling C, Duschl A. Both interleukin 4 and interleukin 13 induce tyrosine phosphorylation of the 140-kDa subunit of the interleukin 4 receptor. J Biol Chem 1995; 270:966-70.
372. Keegan AD, Johnston JA, Tortolani PJ et al. Similarities and differences in signal transduction by interleukin 4 and interleukin 13: analysis of Janus kinase activation. Proc Natl Acad Sci USA 1995; 92:7681-5.
373. Finbloom DS, Winestock KD. IL-10 induces the tyrosine phosphorylation of tyk2 and Jak1 and the differential assembly of STAT1 alpha and STAT3 complexes in human T cells and monocytes. J Immunol 1995; 155:1079-90.
374. Igarashi K-I, Garotta G, Ozmen L et al. Interferon-gamma induces tyrosine phosphorylation of interferon-gamma receptor and regulated association of protein tyrosine kinases, Jak1 and Jak2, with its receptor. J Biol Chem 1994; 269:14333-6.
375. Tanner JW, Chen W, Young RL et al. The conserved box 1 motif of cytokine receptors is required for association with JAK kinases. J Biol Chem 1995; 270:6523-30.
376. Lebrun JJ, Ali S, Ullrich A et al. Proline-rich sequence-mediated Jak2 association to the prolactin receptor is required but not sufficient for signal transduction. J Biol Chem 1995; 270:10664-70.
377. Pawson T, Gish GD. SH2 and SH3 domains: from structure to function. Cell 1992; 71:359-62.

378. Pawson T. SH2 and SH3 domains in signal transduction. Adv Cancer Res 1994; 64:87-110.
379. Farrar MA, Fernandez Luna J, Schreiber RD. Identification of two regions within the cytoplasmic domain of the human interferon-gamma receptor required for function. J Biol Chem 1991; 266:19626-35.
380. Myers MG, Jr., Wang LM, Sun XJ et al. Role of IRS-1-GRB-2 complexes in insulin signaling. Mol Cell Biol 1994; 14:3577-87.
381. Souza SC, Frick GP, Yip R et al. Growth hormone stimulates tyrosine phosphorylation of insulin receptor substrate-1. J Biol Chem 1994; 269:30085-8.
382. Argetsinger LS, Hsu GW, Myers MG Jr et al. Growth hormone, interferon-gamma, and leukemia inhibitory factor promoted tyrosyl phosphorylation of insulin receptor substrate-1. J Biol Chem 1995; 270:14685-92.
383. Sun XJ, Wang LM, Zhang Y et al. Role of IRS-2 in insulin and cytokine signaling. Nature 1995; 377:173-7.
384. Wang LM, Myers MG Jr, Sun XJ et al. IRS-1: essential for insulin- and IL-4-stimulated mitogenesis in hematopoietic cells. Science 1993; 261:1591-4.
385. Wang YD, Wong K, Wood WI. Intracellular tyrosine residues of the human growth hormone receptor are not required for the signaling of proliferation or Jak-STAT activation. J Biol Chem 1995; 270:7021-4.
386. Nice E, Layton J, Fabri L et al. Mapping of the antibody- and receptor-binding domains of granulocyte colony-stimulating factor using an optical biosensor. Comparision with enzyme-linked immunosorbent assay competition studies. Journal of Chromatography 1993; 646:159-68.
387. Quelle FW, Thierfelder W, Witthuhn BA et al. Phosphorylation and activation of the DNA binding activity of purified Stat1 by the Janus protein-tyrosine kinases and the epidermal growth factor receptor. J Biol Chem 1995; 270:20775-80.
388. Novak U, Harpur AG, Paradiso L et al. Colony-stimulating factor 1-induced STAT1 and STAT3 activation is accompanied by phosphorylation of Tyk2 in macrophages and Tyk2 and JAK1 in fibroblasts. Blood 1995; 86:2948-56.
389. Vignais M-L, Sadowski HB, Watling D et al. Platelet-derived growth factor induces phosphorylation of multiple JAK family kinases and STAT proteins. Mol Cell Biol 1996; 16:1759-69.
390. Marrero MB, Schieffer B, Paxton WG et al. Direct stimulation of JAK-STAT pathway by the angiotensin II AT1 receptor. Nature 1995; 375:247-50.

391. Bhat GJ, Thekkumkara TJ, Thomas WG et al. Angiotensin II stimulates sis-inducing factor-like DNA binding activity. Evidence that the AT1A receptor activates transcription factor-Stat91 and/or a related protein. J Biol Chem 1994; 269:31443-9.
392. Bhat GJ, Thekkumkara TJ, Thomas WG et al. Activation of the STAT pathway by angiotensin II in T3CHO/AT1A cells. Cross-talk between angiotensin II and interleukin-6 nuclear signaling. J Biol Chem 1995; 270:19059-65.
393. Lutticken C, Coffer P, Yuan J et al. Interleukin-6-induced serine phosphorylation of transcription factor APRF: evidence for a role in interleukin-6 target gene induction. FEBS Lett 1995; 360:137-43.
394. Baumann H, Won KA, Jahreis GP. Human hepatocyte-stimulating factor-III and interleukin-6 are structurally and immunologically distinct but regulate the production of the same acute phase plasma proteins. J Biol Chem 1989; 264:8046-51.
395. Wakao H, Gouilleux F, Groner B. Mammary gland factor (MGF) is a novel member of the cytokine regulated transcription factor gene family and confers the prolactin response. EMBO J 1995; 14:854-5.
396. Kessler DS, Veals SA, Fu X-Y et al. Interferon alpha regulates nuclear translocation and DNA-binding affinity of ISGF3, a multimeric transcriptional activator. Genes Dev 1990; 4:1753-65.
397. Lamb P, Seidel HM, Haslam J et al. STAT protein complexes activated by interferon-gamma and gp130 signaling molecules differ in their sequence preferences and transcriptional induction properties. Nucleic Acids Res 1995; 23:3283-9.
398. Sims SH, Cha Y, Romine MF et al. A novel interferon-inducible domain: structural and functional analysis of the human interferon regulatory factor 1 gene promoter. Mol Cell Biol 1993; 13:690-702.
399. Haque SJ, Williams BR. Identification and characterization of an interferon (IFN)-stimulated response element-IFN-stimulated gene factor 3-independent signaling pathway for IFN-alpha. J Biol Chem 1994; 269:19523-9.
400. Guyer NB, Severns CW, Wong P et al. IFN-gamma induces a p91/Stat1 alpha-related transcription factor with distinct activation and binding properties. J Immunol 1995; 155:3472-80.
401. Pearse RN, Feinman R, Ravetch JV. Characterization of the promoter of the human gene encoding the high-affinity IgG receptor: transcriptional induction by gamma-interferon is mediated through common DNA response elements. Proc Natl Acad Sci USA 1991; 88:11305-9.
402. Perez C, Wietzerbin J, Benech PD. Two cis-DNA elements involved in myeloid-cell-specific expression and gamma interferon (IFN-gamma) activation of the human high-affinity Fc gamma re-

ceptor gene: a novel IFN regulatory mechanism. Mol Cell Biol 1993; 13:2182-92.
403. Klingmuller U, Lorenz U, Cantley LC et al. Specific recruitment of SH-PTP1 to the erythropoietin receptor causes inactivation of JAK2 and termination of proliferative signals. Cell 1995; 80:729-38.
404. Tsui HW, Siminovitch KA, de Souza L et al. Motheaten and viable motheaten mice have mutations in the haematopoietic cell phosphatase gene. Nat Genet 1993; 4:124-9.
405. David M, Chen HE, Goelz S et al. Differential regulation of the alpha/beta interferon-stimulated JAK-STAT pathway by the SH2 domain-containing tyrosine phosphatase SHPTP1. Mol Cell Biol 1995; 15:7050-8.
406. Petricoin III E, David M, Igarashi K et al. Inhibition of alpha but not gamma interferon signal transduction by phorbol esters is mediated by a tyrosine phosphatase. Mol Cell Biol 1996; 16:1419-24.
407. Simon MA, Bowtell DD, Dodson GS et al. Ras1 and a putative guanine nucleotide exchange factor perform crucial steps in signaling by the sevenless protein tyrosine kinase. Cell 1991; 67:701-16.
408. Simon MA, Carthew RW, Fortini ME et al. Signal transduction pathway initiated by activation of the sevenless tyrosine kinase receptor. Cold Spring Harb Symp Quant Biol 1992; 57:375-80.
409. Aroian RV, Koga M, Mendel JE et al. The let-23 gene necessary for Caenorhabditis elegans vulval induction encodes a tyrosine kinase of the EGF receptor subfamily [see comments]. Nature 1990; 348:693-9.
410. Han M, Sternberg PW. let-60, a gene that specifies cell fates during C. elegans vulval induction, encodes a ras protein. Cell 1990; 63:921-31.
411. Sternberg PW, Horvitz HR. Signal transduction during C. elegans vulval induction. Trends Genet 1991; 7:366-71.
412. Gutjahr T, Frei E, Noll M. Complex regulation of early paired expression: initial activation by gap genes and pattern modulation by pair-rule genes. Development 1993; 117:609-23.
413. Jackle H, Hoch M, Pankratz MJ et al. Transcriptional control by Drosophila gap genes. J Cell Sci Suppl 1992; 16:39-51.
414. Hulskamp M, Tautz D. Gap genes and gradients—the logic behind the gaps. Bioessays 1991; 13:261-8.
415. Berleth T, Burri M, Thoma G et al. The role of localization of bicoid RNA in organizing the anterior pattern of the Drosophila embryo. EMBO J 1988; 7:1749-56.
416. Driever W, Nusslein Volhard C. The bicoid protein determines position in the Drosophila embryo in a concentration-dependent manner. Cell 1988; 54:95-104.

417. Gaul U, Jackle H. Role of gap genes in early Drosophila development. Adv Genet 1990; 27:239-75.
418. Perrimon N, Mahowald AP. l(1)hopscotch, A larval-pupal zygotic lethal with a specific maternal effect on segmentation in Drosophila. Dev Biol 1986; 118:28-41.
419. Harrison DA, Binari R, Nahreini TS et al. Activation of a Drosophila Janus kinase (JAK) causes hematopoietic neoplasia and developmental defects. EMBO J 1995; 14:2857-65.
420. Hanratty WP, Dearolf CR. The Drosophila *Tumorous-lethal* hematopoietic oncogene is a dominant mutation in the *hopscotch* locus. Mol Gen Genet 1993; 238:33-7.
421. Rizki RM, Rizki TM. Cell interactions in the differentiation of a melanotic tumor in Drosophila. Differentiation 1979; 12:167-78.
422. Hanratty WP, Ryerse JS. A genetic melanotic neoplasm of Drosophila melanogaster. Dev Biol 1981; 83:238-49.
423. Corwin HO, Hanratty WP. Characterization of a unique lethal tumorous mutation in Drosophila. Mol Gen Genet 1976; 144:345-7.
424. Nappi AJ, Silvers M. Cell surface changes associated with cellular immune reactions in Drosophila. Science 1984; 225:1166-8.
425. Luo H, Hanratty WP, Dearolf CR. An amino acid substitution in the Drosophila hopTum-l Jak kinase causes leukemia-like hematopoietic defects. EMBO J 1995; 14:1412-20.
426. Hou XS, Melnick MB, Perrimon N. *marelle* acts downstream of the *Drosophila* HOP/JAK kinase and encodes a protein similar to the mammalian STATs. Cell 1996; 84:411-9.
427. Yan R, Small S, Desplan C et al. Identification of the STAT gene that functions in Drosophila development. Cell 1996; 84:421-30.
428. Zinyk DL, McGonnigal BG, Dearolf CR. Drosophila awdK-pn, a homologue of the metastasis suppressor gene nm23, suppresses the Tum-1 haematopoietic oncogene. Nat Genet 1993; 4:195-201.
429. Teng DH, Engele CM, Venkatesh TR. A product of the prune locus of Drosophila is similar to mammalian GTPase-activating protein [see comments]. Nature 1991; 353:437-40.
430. Teng DH, Bender LB, Engele CM et al. Isolation and characterization of the prune locus of Drosophila melanogaster. Genetics 1991; 128:373-80.
431. Barnes TM, Burglin TR. Prune function? [letter; comment]. Nature 1992; 355:504-5.
432. Biggs J, Tripoulas N, Hersperger E et al. Analysis of the lethal interaction between the prune and killer of prune mutations of Drosophila. Genes Dev 1988; 2:1333-43.
433. Biggs J, Hersperger E, Steeg PS et al. A Drosophila gene that is homologous to a mammalian gene associated with tumor metastasis codes for a nucleoside diphosphate kinase. Cell 1990; 63:933-40.

434. Lascu I, Chaffotte A, Limbourg Bouchon B et al. A Pro/Ser substitution in nucleoside diphosphate kinase of Drosophila melanogaster (mutation killer of prune) affects stability but not catalytic efficiency of the enzyme. J Biol Chem 1992; 267:12775-81.
435. Hackstein JH. The lethal prune/killer-of-prune interaction of Drosophila causes a syndrome resembling human neurofibromatosis (NF1). Eur J Cell Biol 1992; 58:429-44.
436. Cutler RL, Liu L, Damen JE et al. Multiple cytokines induce the tyrosine phosphorylation of Shc and its association with Grb2 in hemopoietic cells. J Biol Chem 1993; 268:21463-5.
437. Carroll MP, Spivak JL, McMahon M et al. Erythropoietin induces Raf-1 activation and Raf-1 is required for erythropoietin-mediated proliferation. J Biol Chem 1991; 266:14964-9.
438. He TC, Zhuang H, Jiang N et al. Association of the p85 regulatory subunit of phosphatidylinositol 3-kinase with an essential erythropoietin receptor subdomain. Blood 1993; 82:3530-8.
439. Damen JE, Mui AL, Puil L et al. Phosphatidylinositol 3-kinase associates, via its Src homology 2 domains, with the activated erythropoietin receptor. Blood 1993; 81:3204-10.
440. Miura O, Nakamura N, Ihle JN et al. Erythropoietin-dependent association of phosphatidylinositol 3-kinase with tyrosine-phosphorylated erythropoietin receptor. J Biol Chem 1994; 269:614-20.
441. Izuhara K, Feldman RA, Greer P et al. Interaction of the *c-fes* proto-oncogene product with the interleukin-4 receptor. J Biol Chem 1994; 269:18623-9.
442. David M, Petricoin E, III, Benjamin C et al. Requirement for MAP kinase (ERK2) activity in interferon α- and interferon β-stimulated gene expression through STAT proteins. Science 1995; 269:1721-3.
443. Migone TS, Lin JX, Cereseto A et al. Constitutively activated Jak-STAT pathway in T cells transformed with HTLV-I. Science 1995; 269:79-81.
444. Yu CL, Meyer DJ, Campbell GS et al. Enhanced DNA-binding activity of a Stat3-related protein in cells transformed by the Src oncoprotein. Science 1995; 269:81-3.
445. Jove R, Hanafusa H. Cell transformation by the viral *src* oncogene. Ann Rev Cell Biol 1987; 3:31-56.
446. Courtneidge SA. Activation of the pp60$^{c\text{-}src}$ kinase by middle T antigen binding or by dephosphorylation. EMBO J 1985; 4:1471-7.
447. Roussel RR, Brodeur SR, Shalloway D et al. Selective binding of activated pp60c-src by an immobilized synthetic phosphopeptide modeled on the carboxyl terminus of pp60c-src. Proc Natl Acad Sci USA 1991; 88:10696-700.
448. Danial NN, Pernis A, Rothman PB. Jak-STAT signaling induced by the *v-abl* oncogene. Science 1995; 269:1875-7.

449. Rosenberg N, Witte ON. The viral and cellular forms of the abelson (*abl*) oncogene. Advances in Virus Research 1988; 35:39-81.
450. Cook WD, Metcalf D, Nicola NA et al. Malignant transformation of a growth factor-dependent myeloid cell line by Ableson Virus without evidence of an autocrine mechanism. Cell 1985; 41:677-83.
451. Kincade PW, Lee G, Pietrangeli CE et al. Cells and molecules that regulate B lymphopoiesis in bone marrow. Annu Rev Immunol 1989; 7:111-43.
452. Yoshida M, Seiki M. Recent advances in the molecular biology of HTLV-1: trans-activation of viral and cellular genes. Annu Rev Immunol 1987; 5:541-59.
453. Yodoi J, Uchiyama T. Diseases associated with HTLV-I: virus, IL-2 receptor dysregulation and redox regulation. Immunol Today 1992; 13:405-11.
454. Merida I, Williamson P, Kuziel WA et al. The serine-rich cytoplasmic domain of the interleukin-2 receptor β chain is essential for interleukin-2-dependent tyrosine protein kinase and phosphatidylinositol-3-kinase activation. J Biol Chem 1993; 268: 6765-70.
455. Russell SM, Tayebi N, Nakajima H et al. Mutation of Jak3 in a patient with SCID: essential role of Jak3 in lymphoid development. Science 1995; 270:797-800.
456. Puck JM. Molecular and genetic basis of X-linked immunodeficiency disorders. J Clin Immunol 1994; 14:81-9.
457. Macchi P, Villa A, Gillani S et al. Mutations of Jak-3 gene in patients with autosomal severe combined immune deficiency (SCID). Nature 1995; 376:65-8.
458. Cao X, Shores EW, Hu Li J et al. Defective lymphoid development in mice lacking expression of the common cytokine receptor gamma chain. Immunity 1995; 2:223-38.
459. DiSanto JP, Muller W, Guy Grand D et al. Lymphoid development in mice with a targeted deletion of the interleukin 2 receptor gamma chain. Proc Natl Acad Sci USA 1995; 92:377-81.
460. Ryder K, Lau LF, Nathans D. A gene activated by growth factors is related to the oncogene v-jun. Proc Natl Acad Sci USA 1988; 85:1487-91.
461. Fields S, Song O. A novel genetic system to detect protein-protein interactions. Nature 1989; 340:245-6.
462. Chevray PM, Nathans D. Protein interaction cloning in yeast: identification of mammalian proteins that react with the leucine zipper of Jun. Proc Natl Acad Sci USA 1992; 89:5789-93.
463. Angel P, Smeal T, Meek J et al. Jun and v-jun contain multiple regions that participate in transcriptional activation in an interdependent manner. New Biol 1989; 1:35-43.

464. Bohmann D, Tjian R. Biochemical analysis of transcriptional activation by Jun: differential activity of c- and v-Jun. Cell 1989; 59:709-17.
465. Capecchi MR. Altering the genome by homologous recombination. Science 1989; 244:1288-92.
466. Thomis DC, Gurniak CB, Tivol E et al. Defects in B lymphocyte maturation and T lymphocyte activation in mice lacking Jak3. Science 1995; 270:794-7.
467. Nosaka T, van Deursen JM, Tripp RA et al. Defective lymphoid development in mice lacking Jak3. Science 1995; 270:800-2.
468. von Freeden Jeffry U, Vieira P, Lucian LA et al. Lymphopenia in interleukin (IL)-7 gene-deleted mice identifies IL-7 as a nonredundant cytokine. J Exp Med 1995; 181:1519-26.
469. Peschon JJ, Morrissey PJ, Grabstein KH et al. Early lymphocyte expansion is severely impaired in interleukin 7 receptor-deficient mice. J Exp Med 1994; 180:1955-60.
470. Stancato LF, David M, Carter-Su C et al. Preassociation of STAT1 with STAT2 and STAT3 in separate signaling complexes prior to cytokine stimulation. J Biol Chem 1996; 271:4134-7.
471. Briscoe J, Rogers NC, Witthuhn BA et al. Kinase-negative mutants of JAK1 can sustain interferon-gamma-inducible gene expression but not an anti-viral state. EMBO Journal 1996; 15:799-809.
472. Guschin D, Rogers N, Briscoe J et al. A major role for the protein tyrosine kinase JAK1 in the JAK-STAT signal transduction pathway in response to interleukin-6. EMBO J 1995; 14:1421-9.
473. Gupta S, Yan H, Wong LH et al. The SH2 domains of STAT1 and STAT2 mediate multiple interactions in the transduction of IFN alpha signals. EMBO Journal 1996; 15:1075-84.
474. Quelle FW, Sato N, Witthuhn BA et al. JAK2 associates with the beta c chain of the receptor for granulocyte-macrophage colony-stimulating factor, and its activation requires the membrane-proximal region. Mol Cell Biol 1994; 14:4335-41.

INDEX

Items in italics denote figures (f) or tables (t).

S

T

GPSR Compliance

The European Union's (EU) General Product Safety Regulation (GPSR) is a set of rules that requires consumer products to be safe and our obligations to ensure this.

If you have any concerns about our products, you can contact us on ProductSafety@springernature.com

In case Publisher is established outside the EU, the EU authorized representative is:

Springer Nature Customer Service Center GmbH
Europaplatz 3
69115 Heidelberg, Germany

Batch number: 10615115

Printed by Printforce, the Netherlands